Project Benefit Realisation and Project Management

Project Benefit Realisation and Project Management

The 6Q Governance Approach

Raymond C. Young, MBA, PhD, FAIPM, FGIA

Vedran Zerjav, PhD

WILEY Blackwell

This edition first published 2022
© 2022 John Wiley & Sons Ltd

The right of Raymond C. Young and Vedran Zerjav to be identified as the authors of this work has been asserted in accordance with law.

Registered Offices
John Wiley & Sons, Inc., 111 River Street, Hoboken, NJ 07030, USA
John Wiley & Sons Ltd, The Atrium, Southern Gate, Chichester, West Sussex, PO19 8SQ, UK

Editorial Office
9600 Garsington Road, Oxford, OX4 2DQ, UK

For details of our global editorial offices, customer services, and more information about Wiley products visit us at www.wiley.com.

Wiley also publishes its books in a variety of electronic formats and by print-on-demand. Some content that appears in standard print versions of this book may not be available in other formats.

Library of Congress Cataloging-in-Publication data applied for

ISBN: 9781119367888

Cover Design: Wiley
Cover Image: © matejmo/Getty

Set in 9.5/12.5pt STIXTwoText by Straive, Pondicherry, India

Printed and bound by CPI Group (UK) Ltd, Croydon, CR0 4YY

C9781119367888_310821

Contents

List of Illustrations

Preface

As we are writing this handbook, the COVID-19 crisis is unfolding, creating challenges and opportunities of unprecedented scale for the economies and societies around the world. A number of projects are having their budgets axed and objectives put on hold but at the same time, many new projects are being initiated to help us all prepare to function and live in the 'new normal'. Project sponsors, investors, and financiers are taking a risk-averse view as to what projects will be needed and what to support in the post-pandemic 'brave new world'. Indeed, we felt that this handbook is a timely response to the challenges of this situation, focusing on project benefits and the design of projects in the light of their benefits, rather than the conventional criteria of project success.

Our ambition is to help organisational project clients and their project executives navigate these uncertain times. The book is deliberately brief and written for Corporate Boards and their 'accidental project sponsors'. Managers do not suddenly acquire the knowledge to govern a big project when they are promoted to senior management. It is far more common to be delegated the role and to become a project sponsor by 'accident'. Once delegated, the common experience is to find most of the advice one is given is not helpful, and success or failure becomes dependent on your instincts as a project sponsor. In the post-pandemic world, projects are more critical to the survival of organisations and we cannot fall back on such a hit-and-miss approach. This handbook addresses this issue by distilling the experience of senior managers and presenting guidance in the form of six key questions illustrated by case studies.

The guidance is informed by decades of research and has been tested against a database of hundreds of projects to confirm its effectiveness. It is also supported by a companion website (www.6qgovernance.com) where readers can read the latest insights and post questions to the authors and their peers to get timely advice on how to govern their projects for the new normal.

Authors CV

Raymond C. Young – Career Summary

Raymond Young is an international authority in project governance. His research has been published by Standards Australia as HB280-2006, a handbook explaining how boards and top managers influence business projects to succeed. His career objective is to help clients realise strategic business benefits from their projects rather than to simply come in on-time on-budget.

Raymond is a Fellow of the Australian Institute of Project Management (FAIPM) and a Fellow of the Governance Institute of Australia (FGIA). He is a founding member of the committee that developed the Australian and international governance standards AS8016 and ISO38500. Raymond's career alternates between industry and academia. He has recently taken up a Senior Associate Professor role at Xi'an Jiaotong-Liverpool University in Suzhou, China after a lengthy period as an academic at both the University of New South Wales and the University of Canberra. He has a decade of management consulting experience culminating in a CIO role within Fujitsu Australia.

Until recently, Raymond has been advising Australian Federal Government agencies on how to improve their project, programme, and portfolio management practices. His other expertise includes performance measurement, business process reengineering, activity-based management, and logistics. His significant clients have included Colgate-Palmolive, BHP, Commonwealth Bank of Australia, Telstra, Department of Health, Electricity Trust of South Australia, and Prospect Electricity.

Qualifications and Accreditations

- Fellow of the Governance Institute of Australia (FGIA)
- Fellow of the Australian Institute of Project Management (FAIPM)
- Graduate of the Australian Institute of Company Directors (GAICD)

- Doctor of Philosophy (PhD), Macquarie Graduate School of Management, Macquarie University, 2006
- Master of Business Administration (MBA), University of Sydney, 1988–1992
- Graduate Diploma in Education (DipEd), Sydney College of Advanced Education, 1986
- Bachelor of Building Science (BBSc), Victoria University of Wellington, New Zealand, 1982–1984

Work History

2019–current	**Senior Associate Professor** Xi'an Jiaotong-Liverpool University
2016–2019	**Senior Lecturer** University of NSW, Canberra
2010–2016	**Assistant Professor** University of Canberra
2008–2010	**Practice Lead, Project Governance** e8 Consulting
2002–2008	**Lecturer** Macquarie University
1994–1998	**CIO** Fujitsu Australia & FBA Computer Technology Services
1991–1994	**Management Consultant** Deloitte Touche Tohmatsu

Vedran Zerjav – Career Summary

Vedran Zerjav is an Associate Professor of Infrastructure Project Management in the Bartlett, UCL. He is a scholar of projects with an interest in a range of organisational issues in project-based organisational forms. His main areas of interest include strategic, operational, and value considerations in projects and his empirical focus is on urban infrastructure and its delivery. He is a qualitative researcher with an interest in hybrid and novel methodologies for project studies. Vedran's engagement with the world of project management practice is extensive and spans research and advisory roles working with major infrastructure clients and professional bodies such as the Association for Project Management and Project Management Institute.

Qualifications

Doctor of Technical Sciences (Dr.techn.), Vienna University of Technology, 2012
Diploma in Civil Engineering (Dipl.ing./M.Eng.), University of Zagreb, 2006

Work History

2014–Present	University College London, Bartlett School of Construction and Project Management
2012–2014	University of Twente (The Netherlands), Department of Construction Management and Engineering
2006–2012	University of Zagreb (Croatia), Faculty of Civil Engineering, Department of Construction Management and Economics

1

Introduction

The Board, Governance and Projects

A corporate board's role is to 'ensure management is focused on above-average returns while taking account of risk' [1, 2]. At a minimum, in our post-COVID world, the board needs to ensure the survival of an organisation. For most organisations, business-as-usual is not an option because the environment has changed so much. Projects have always been important but now they are even more so because projects are the vehicle to take us from where we are now to where we need to be to survive *and* thrive. Project failure is not an option and projects are more of a boardroom issue than they ever were.

However, at the board and senior management level, we have tended to think of projects as someone else's problem. Boards willingly take oversight of a company's strategy but they seldom follow through on the insight that strategy [3] is implemented through projects. When we pause, reflect, and examine the success rates of strategy and projects, all the evidence suggests there is a very large strategy to performance gap [4]. Fewer than 10% of strategies are fully implemented [5], most large projects fail to live up to expectations [6] and between half to two-thirds of projects either fail outright or deliver no discernible benefits [7]. There is a major deficiency in practice. Projects are troublesome with high capital costs, long time frames, and difficulties in delivery. Apart from the funding decision, most of us would prefer not to have to deal with projects at all. However, the world has changed and now, unless projects succeed, there may be no business at all. Boards and their advisors need to discuss projects in terms of the strategic benefits to be realised and go beyond the traditional focus on time and cost.

Project results appear to be no better in the public sector where hundreds of billions of dollars are invested annually in projects that contribute little to policy goals [8, 9]. If this pattern were to continue into the post-COVID-world: huge sums

Project Benefit Realisation and Project Management: The 6Q Governance Approach,
First Edition. Raymond C. Young and Vedran Zerjav.
© 2022 John Wiley & Sons Ltd. Published 2022 by John Wiley & Sons Ltd.

of money will be wasted, more people will die, and the economy may take decades longer than necessary to recover. At the senior management level, we need to go beyond issues of probity and simply doing it right. We need to learn to focus on the more important issue of realising strategic benefits in the right projects.

This book has been written to address the needs of the board and senior management. It deals with projects, but it is not a project management book. Instead, it focuses on implementing strategy, policy, and creating value through projects.

This book has been written because boards and top managers need better guidance. There is a paradox in project management: Project management is a mature discipline but following the standard guidelines does not automatically lead to success [10, 11]. There is widespread confusion [12] between project management success (on-time on-budget) and project success (realisation of business benefits) and most project management books incorrectly imply that one will lead to another. The guidelines that exist do not explicitly acknowledge that it is more important to realise strategic benefits than to simply come in on-time and on-budget. This notion is particularly important in the world of the new normal where it will be even less important to focus on time and budget criteria of projects and focus on achieving strategic goals of organisations. A classic example occurred during the 2008 Global Financial Crisis when the Australian government spent billions of dollars on school buildings to keep the economy moving. Schools got a new hall whether they needed one or not and the public dialogue was all about time and cost. The expenditure helped the economy, but no one thought to question whether equally important strategic educational goals such as literacy and numeracy had improved because of the projects. We risk repeating the same mistake as we commission projects to overcome the downturn in the economy caused by COVID-19.

Projects rarely succeed in realising their expected benefits without the top management support [13, 14] and this is even more the case in the post-pandemic world. Project management books provide little to no guidance for the senior management and, as a result, top managers are often not sure how to govern their projects to succeed. This insight informed the development of an international standard ISO38500 [15] and is based on HB280 [16]: an Australian Handbook for boards and their senior managers on how to govern ICT projects to succeed. HB280 is a research report distilling the experience of senior managers to identify what they did right and what they did wrong in governing their projects to succeed (or fail). In the words of one senior manager, 'These big projects are a little like marriage, you don't do them often enough to get practiced at it [and it is valuable to debrief to identify what you did to cause them to succeed]'.

HB280 presented five cases and distilled the lessons learned into six key questions that boards and their delegates should be asking when governing projects. These six questions were trademarked as 6Q Governance™ to make them more memorable. The 6Q Governance Framework has been tested with an international dataset to prove it works with all types of projects [17]. We see an adaptation of this framework as particularly well suited for the current transition into the new normal where projects need to be constantly revisited and interrogated by their clients and sponsors rather than have their requirements frozen in time and driven to completion with the least disruption. This handbook will show boards and their senior managers how to apply the 6Q Governance Framework by asking just six key questions at different stages in a project lifecycle. This is the first handbook to present projects in the context of the needs of the top manager and is written for board members, their 'accidental' project sponsors, the business project manager, and their project advisors.

A Diagnostic Toolbox for Project Executives

Our aim in this handbook is to set forward the basic diagnostic toolbox that will help executives understand the strategic health of their project portfolio and to know whether they are on track to realise the benefits that they were originally set out to achieve. We are offering the equivalent of a doctor's stethoscope and a diagnostic map: six places to check to determine the health of your project. As in any medical check-up, the general condition will depend on how several forces interact with each other in a systemic way. If a doctor requests a blood test for example, what we will get is a few measurements with values within certain ranges that are deemed as normal for any given parameter. If the blood counts are all within the expected ranges, no action is to be taken. If, however, a certain value is found to be outside the 'normal' range, then action needs to be taken to explore the nature of the 'disease'.

Our method establishes dialogue and questioning as the main mode of engagement to achieve alignment between projects, strategy, and policy and, thus, undertake a project health check. The diagnostics will allow us to probe into different aspects of the project in order to guide reflection on the different options for the project business case and if and how it is going to fulfil its objectives. These questions are the backbone of our thinking in this book and they are the equivalent of the diagnostic map shown in Figure 1.1. Armed with the knowledge of where to look and listen, you are the stethoscope to bring your projects into strategic conversations.

We next introduce a toolbox grouped by six questions that you can use to make sense of any given number of projects in their organisation and then govern at the right time. These six questions were derived by rigorously looking at both successful and unsuccessful projects and asking what could have been done to improve the business outcome. The 6Q Governance business canvas is given below and each of its areas will be elaborated in more detail in Chapter 2 of this book.

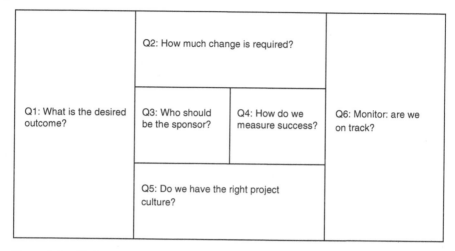

Figure 1.1 6Q Governance (TM) as a business canvas.

The rest of the handbook is organised in the following way:

- Key concepts
- Chapter 2: The 6Q Governance questions
- Chapter 3: Tools and techniques
- Chapter 4: Further insights
- Appendices: Detailed case studies – for practice

Key Concepts

Project Management Success vs. Project Success

Before we launch into the six Questions, there is an important concept to establish: what is success in the context of the new normal? A lot has been written about project success in terms of time and cost, so much so that for most people, this is their only understanding of success. It is a blind spot for most practitioners. As long as the project delivery team can argue that they delivered what was agreed in the project brief and as long as they can do it within the constraints of time and budget, they will be seen as successful. Of course, there are many nuances to this argument by adding additional dimensions such as innovation, stakeholder management, leadership, entrepreneurship, and others but, ultimately, they all boil down to whether the project was on-time and on-budget.

This is the idea of project success that we are criticising in this book. We are not saying that projects should be seen as if they are unbounded by earthly constraints such as time and budget. But we are saying that a more strategic view of project success is necessary to come up with a meaningful execution strategy. For example, the question of whether stopping a failing project early should be considered a failure or success is never asked in the traditional execution sense. It would invariably be considered a failed project. In the same vein, a project that delivered nothing but a white elephant – an asset that is expensive to maintain but of very little or no use at all – would be considered a success as long as its deliverable remotely resembles what was promised in the brief when it was put together even if the world may have changed since then – and it will have changed.

So, how can we then change the emphasis to project success rather than project management success? This is the main riddle we will try to address with this handbook. Having been involved with projects and their management in various shapes and forms collectively for several decades, the main problem we've seen is one of failing to achieve alignment between policy, strategy, operations (outcome thinking), and project delivery (output thinking).

Currently, projects are not measured on how they are aligned with the ever-evolving strategy and operational requirements of their respective organisations, but on how they aligned with their own plans suggested at the outset. Therefore, the front-end planning ends up being a proxy for measuring project success, regardless of whether the project plans made sense in the first place. Then, as the project unfolds, project plans are often put forward with an entirely different mindset and agenda than what is described in the business case. Indeed, the underlying rationale of the project brief is to provide an early outline and a justification for the project with the main ambition of going very little beyond having the sponsor pushing the go button to get the project sanctioned. In this handbook, we will argue that this is the problem to be solved. At the heart of the problem are the often-rushed planning and design decisions that take place at the inception of the project. This is what needs fixing, the fact that project execution does not conform to them is simply a corollary.

We should embrace serendipity, but not forget what projects are for – getting things done. But 'things' can often be a lot different to that what we think at the project outset and 'done' can mean different things in different situations. We are writing this handbook to understand the success of projects as its alignment with not be on time, on cost, outputs, but rather with the operational and strategic purpose that the organisation has set out.

This change in emphasis is exactly what makes it nightmarishly difficult to deliver the project as an 'accidental project sponsor' as any one of us can potentially find ourselves in such an accidental role. We have started by putting this marker on the ground: projects need to focus on realising (evolving) strategic goals. Project management success (on-time on-budget) is a secondary objective, and the so-called experts must not be allowed to dominate the conversation by their misguided efforts. This book will try to shed some light and build a roadmap for what is clearly a daunting journey full of trials and difficulties. But before we start, let us take one more deep breath and discuss strategy and its importance for the success of projects.

Strategy and Policy Execution

After discussing what projects are for strategy and business operations, it is time to reflect on the other side of the coin – what strategy and business operations are for the projects we deliver. The short answer is – everything. Let us start by discussing the importance of strategy.

We would like you to pause and rank the importance of the following governance issues:

- Risk and compliance – Avoid litigation, reputational risk
- Strategy and planning
- Board composition, diversity, and performance
- Government regulation
- Executive compensation – alignment with shareholder expectations of performance
- New technology and its impact on the business

Strategy is probably far more important than you think. At a minimum strategy ensures the long-term survival of an organisation and good strategy will lead to above-average performance. The following statistics show that these are not motherhood statements:

- For small businesses, ineffective strategy contributes to the problem of 50% surviving no longer than 5 years and 64% not surviving for more than 10 years [18].
- Poor strategy in large businesses results in underperformance. Booz Allen Hamilton found in a five-year study of under-performing US organisations [19] that 60% of the value destroyed was due to strategic errors, 27% due to operational errors, and 13% due to compliance problems.
- In the public sector, strategy is a confused concept [20] and we often talk about policy instead. A study of the State of Victoria in Australia, normally considered an exemplar, found $100B had been invested into projects over a 10-year period without any evidence any high-level policy goals had improved [9]. A follow-up study in the State of NSW also in Australia suggested that, more generally, only one in five policy goals are positively impacted [8].

In discussing the role of a board, Henry Bosch (former chair of the Australian National Companies and Securities Commission, now ASIC) states:

> The board's first responsibility is to ensure that the organisation has clearly established goals; objectives and strategies for achieving them; that they are appropriate in the circumstances; and that they are understood by management (1995:93).

Before reading on, we suggest you go back to the earlier question and reconsider the relative importance of the governance issues. On reflection, do you find the strategy is more important than you first thought? Management consultants Booz Allen support a performance rather than a risk emphasis for governance and made the following observation around the time of the Enron, WorldCom, and Tyco scandals:

> More strategic value has been destroyed in the past five years as a result of strategic mismanagement and poor execution . . . than was lost in all of the recent compliance scandals combined [19].

So, strategy is important. However, it is not necessarily the glamorous function that only highly paid consultants can deliver. Ever since seminal works by strategy scholars such as Henry Mintzberg, strategy has been seen as a lot less as clever planning and a lot more as juggling a number of competing priorities and conflicting goals in face of execution and operational challenges. Indeed, strategy and strategic management turn out to be a mundane and organic effort that unfolds in day-to-day activity.

Strategy is increasingly implemented through projects. Ideally, these projects are initiated only after a strategy has been determined. However, in practice, this is not always possible. There are many times when the strategy needs to change and in-flight projects need to be reconsidered to respond to the environment, e.g. changes in legislation. There are other times when action must be taken before one's strategy is fully thought through, e.g. natural disasters. This reactive mode may even be the norm in today's turbulent times as shown by events such as the Global Financial Crisis, globalisation, rapidly changing consumer demand and the COVID-19 pandemic. It is also worth noting that even in the ideal situation when the strategy is carefully thought out beforehand, the strategy is often poorly implemented with some leading consultants such as the Boston Consulting Group and McKinsey admitting (off the record) that as few as 10% of strategies are ever implemented effectively [5]. In all these cases, strategy formulation is not only performed separately but is actively reconsidered as an ongoing part of the implementation. This is quite a different concept to the current practice where top managers engage actively in strategy formulation but seldom consider projects to be a matter of direct concern [21].

Consider this: Boards approve around 40% of all projects [22] and the management of these large-scale expenditures is a fiduciary duty requiring careful oversight. However, Deloitte warns that current practice is 'tantamount to negligence'.

- 50–80% of the time projects do not deliver the expected benefits? [7]
- 29–46% of the time ICT projects are approved with either inadequate or no information? [23]

Nobel Laureate Daniel Kahneman reported statistics that suggest the problem is widespread with ¾ of mergers and acquisitions never paying off, most large capital projects failing to live up to expectations, the majority of efforts to enter new markets abandoned in a few years and 70% of new manufacturing plants closed in their first decade [6]. We believe it is only a matter of time before boards and their management teams are held to account for these lapses in corporate governance.

The point we are making is that it is dysfunctional for top managers to be allowed to fully delegate their responsibility because projects must succeed for an organisation's strategies to be implemented. The evidence is now clear that top management support is the most critical factor for project success [24]. However, it is neither practical nor desirable for top managers to be overly hands-on at the project level. The key is to get the top management feedback at the right time through the project governance process. Boards and their delegates need to know how to 'steer' their projects to success [16].

Are We Talking About Strategy or Policy?

Having extensively discussed the role of strategy for projects, we should mention the important role that policy plays in project execution. Some projects are simply there to enact or enforce a particular public policy. Like the previous argument about the failure of the strategy to achieve business objectives through projects, the situation with policy is no different. This is where megaprojects and infrastructure can teach us a lesson. Generally, the larger and more complex a project becomes, the fuzzier will be its front-end and the less likely that the goals will be met.

Only think about the pilot project of introducing the universal credit system in the UK. This much-awaited incentive has been planned for years to simplify the system of benefits in the UK. Effectively, the new Universal Credit System was to replace six separate income-related benefits and has been piloted since 2013. Nonetheless, the rollout of the system has been facing numerous delays and controversy due to the high complexity and interdependency of the elements that the project is supposed to integrate. As a result, the policy implementation as the main goal of the project has also been hampered.

Clearly, a policy is key to the existence and implementation of such an extensive programme. The other example is the famously failed NHS IT programme that existed between 2005 and 2013 until it was suspended. In that time, the costs skyrocketed from the original estimate of £2.3B to, according to some estimates, around £20B.

Over the last several decades, London has been home to a number of groundbreaking projects which are seen as game-changers for urban planning and policy. Take the example of the east London regeneration plan that was one of the original aims for the London Olympics 2012. One of the programmes was the East London line which contributed to the regeneration of an otherwise deprived area. Its purpose was to improve connectivity for other benefits to happen: job creation, housing, and economic growth.

The other example is Crossrail, a high-speed underground system connecting London's West with the East. The drive was the increasing unaffordability of housing and enabling better connectivity to central London which is home to most businesses. The argument went that if people are better able to travel from suburban zones to Central London, it would have contributed to the productivity of the service sectors (e.g. finance) located in central London areas. Whilst Crossrail similarly faces a number of issues including a delay and budget overrun, any large-scale infrastructure project creates a large number of externalities, both positive and negative, and that those are hard to pinpoint at the front end. The other thing is that those projects are often delivered in the form of megaprojects that are more expensive than £1B. This means that for some of those projects, they can take up substantial chunks of the national GDP. However, as they get bigger, the literature teaches us that they get more fragile and they fail more easily. They are also more difficult to stop once they turn pear-shaped.

There is an entire stream of work on such megaprojects and they are increasingly becoming an area of not only scholarly inquiry but also of governance and management expertise. Studies looking at the success of those projects suggest none have succeeded. In other words, virtually all large-scale infrastructure projects fail to deliver on their expectations [25]. This is arguably because of the interests of planners, engineers, and politicians to move their pet projects rather than ones that are going to bring about real benefits. This leads to projects getting approved – but the ones which look best on paper, not in reality. And they can only look best on paper if someone either 'strategically misrepresented' them or was optimistically biased [26]. The result is that we have the worst projects and the 'survival of the unfittest'. They are chosen based on the agenda that politicians want to get re-elected, they want to be seen as the facilitators of those projects with vast social impact. The Heathrow Airport T5 opening back in 2008 was a good example. Whilst the project was and still is widely considered a success, its chaotic opening made the headlines of the press worldwide. To draw a line under the argument, our stance here would not be as pessimistic. We do not agree that it is the worst projects that get built (indeed there is no evidence for that) but that random infrastructure projects get built at best. However, through their governance structures and a set of local and historical circumstances, they can evolve through time and become something very different from what they were envisioned to be.

So, what is the lesson to be learned for executives here looking at infrastructure and megaprojects? We surely are not trying to teach you here how to govern a £1B+ beast, but we do want you to learn from the experience of trying to tame such beasts. There are a few things we can induce here. It is that projects are more fragile, the more they grow and more complex they become [27]. The second thing is that they evolve with time. The third point is that no matter how hard you try to nail down the specs, it will change. The fourth point is that the project will not only have its immediate benefits but there will be wider externalities that are captured beyond the boundary of the projects. So, what is the solution learning from infrastructure implementation programmes? The short answer is a lot. Most importantly, it allows us to say that governance structures can be designed upfront. We can design our projects by looking at the organisation that delivers them and their alignment between the strategy, policy, and operations.

Design Thinking for Project Governance

The final key concept we want to touch on briefly is design thinking. We believe the role of the accidental project sponsor is as a reflective practitioner [28] or a painter who engages with the canvas in a reflective conversation and by so doing, creates something that has not existed before – the artificial [29]. In this sense, the argument of this book will be based on the idea that projects are designed in an interactive and stepwise manner rather than implemented based on a pre-existing plan. In our experience, we have seen many a project being implemented for no real strategic purpose in terms of benefits realisation. This was always wrong, and it will be even more wrong in the post-2020 pandemic world of projects where investments in any capacity upgrades, assets, or product developments are likely to be scrutinised even more. It is much more likely that there will be more effort in the front end of the project where strategic decisions are made to decide about which project to sanction and equally to set up its governance and organisational structure. For this reason, it is likely that we will need more flexibility and agility in the project development process and that we will have portfolios comprising larger numbers of smaller projects brought together in a modular fashion.

There are two main theoretical points we would like to drive home here. And that is that (i) projects are designed as solutions to policy or business problems. And that (ii) these problems are only seldom straightforward and well-structured. More often than not, they would be what we call 'wicked' [30]. Wicked not in the sense of evil and morally wrong but wicked in the sense of playfully mischievous. The term 'wicked problem' was originally coined in the context of town planning but was then expanded to include all sorts of political, economic, and social problem-solving. Some of the most obvious differences between tame and wicked problems are that the former are describable and determinate, whilst in the latter, problem definition incorporates solution and only indeterminate solutions can be achieved. Design, on the other hand, can be seen as a combination of framing and problem-solving [31] and so can be project management [32].

This book will draw on the main principles of design and design thinking and apply them to the context of projects as they are likely to be needed rather than the traditional top-down mindset focused on on-time and to budget. The main premise is that we think of projects as solutions to problems that need to be designed rather than taken for granted as is sometimes the case.

References

1 Hilmer, F.G. (1993). *Strictly Boardroom: Improving Governance to Enhance Company Performance*. Melbourne: The Business Library.

2 Cadbury (1992). *The Financial Aspects of Corporate Governance*. London: Gee.

3 Kwak, Y.H. and Anbari, F.T. (2009). Analyzing project management research: perspectives from top management journals. *Int. J. Proj. Manag.* 27 (5): 435–446.

4 Mankins, M.C. and Steele, R. (2005). Turning great strategy into great performance. *Harv. Bus. Rev.* 83 (7–8): 64.

5 Kiechell, W. (2010). *The Lords of Strategy*. Boston: Harvard Business Press.

6 Lovallo, D. and Kahneman, D. (2003). Delusions of success: how optimism undermines executive's decisions. *Harv. Bus. Rev.* 81 (7): 56–63.

7 Young, R. (2006). What is the ROI for IT Project Governance? Establishing a benchmark. *2006 IT Governance International Conference*, Auckland, New Zealand (November 2006).

8 Young, R. and Grant, J. (2015). Is strategy implemented by projects? Disturbing evidence in the state of NSW. *Int. J. Proj. Manag.* 33 (1): 15–28.

9 Young, R., Young, M., Jordan, E., and O'Connor, P. (2012). Is strategy being implemented through projects? Contrary evidence from a leader in New Public Management. *Int. J. Proj. Manag.* 30 (8): 887 900.

10 Starkweather, J.A. and Stevenson, D.H. (2011). PMP® certification as a core competency: necessary but not sufficient. *Proj. Manag. J.* 42 (1): 31–41.

11 Joseph, N. and Marnewick, C. (2018). Investing in project management certification: do organisations get their money's worth? *Inf. Technol. Manag.* 19 (1): 51–74.

12 Baccarini, D. (1999). The logical framework for defining project success. *Proj. Manag. J.* 30 (4): 25–32.

13 Young, R. and Poon, S. (2013). Top management support-almost always necessary and sometimes sufficient for success: findings from a fuzzy set analysis. *Int. J. Proj. Manag.* 31 (7): 943–957.

14 Young, R. and Jordan, E. (2008). Top management support: mantra or necessity? *Int. J. Proj. Manag.* 26 (7): 713–725.

15 ISO 38500 (2008). *Corporate Governance of Information Technology*. International Standards Organisation.

16 Standards Australia (2006). *HB280 Case Studies – How Boards and Senior Management Have Governed ICT Projects to Succeed (or Fail)*. Sydney: Standards Australia.

17 Young, R., Chen, W., Quazi, A. et al. (2019). The relationship between project governance mechanisms and project success? An international dataset. *Int. J. Manag. Proj. Bus.* 13 (7): 1496–1521.

18 Watson, J. and Everett, J.E. (1996). Do small businesses have high failure rates? *J. Small Bus. Manag.* 34 (4): 45.

19 Booz Allen Hamilton. Too Much SOX Can Kill You Resolving the Compliance Paradox [Internet]. 2004 [cited 2016 Aug 9]. Available from: https://www.boozallen.com/content/dam/boozallen/media/file/143161.pdf

20 Stewart, J. (2004). The meaning of strategy in the public sector. *Aust. J. Public Adm.* 63 (4): 16–21.

21 Crawford, L. (2005). Senior management perceptions of project management competence. *Int. J. Proj. Manag.* 23 (1): 7–16.

22 KPMG (2005). *Global IT Project Management Survey: How Committed Are You?* KPMG.

23 Deloitte (2007). *What the Board Needs to Know About IT: Phase II Findings: Maximizing Performance Through IT Strategy.* Deloitte LLP.

24 Young, R. and Poon, S. (2013). TOP MANAGEMENT SUPPORT – almost always NECESSARY AND sometimes SUFFICIENT for success: findings from a fuzzy set analysis. *Int. J. Proj. Manag.* 31: 943–957.

25 Flyvbjerg, B. and Sunstein, C.R. (2016). The principle of the malevolent hiding hand; or, the planning fallacy writ large. *Soc. Res. (New. York).* 83 (4): 979–1004.

26 Flyvbjerg, B. (2014). What you should know about megaprojects and why: an overview. *Proj. Manag. J.* 45 (2): 6–19.

27 Ansar, A., Flyvbjerg, B., Budzier, A., and Lunn, D. (2016). Does infrastructure investment lead to economic growth or economic fragility? Evidence from China. *Oxford Rev. Econ. Policy* 32 (3): 360–390.

28 Schön, D.A. (1983). *The Reflective Practitioner: How Professionals Think in Action.* London: Temple Smith.

29 Simon, H.A. (1976). *Administrative Behaviour*, 3e. New York: The Free Press.

30 Rittel, H.W.J. and Webber, M.M. (1973). Dilemmas in a general theory of planning. *Policy Sci.* 4: 155–169.

31 Dorst, K. (2011). The core of 'design thinking'and its application. *Des. Stud.* 32 (6): 521–532.

32 Zerjav, V., Hartmann, T., and Achammer, C. (2013). Managing the process of interdisciplinary design: identifying, enforcing, and anticipating decision-making frames. *Archit. Eng. Des. Manag.* 9 (2): 121–133.

2

How to Govern Projects: Six Questions

Q1. What is the Desired Outcome?

Effective governance of projects requires clarity on the desired outcome and the first 6Q Governance™ question, Q1, is to clarify the link between business outcomes, benefits, and strategy. Research suggests that projects are funded without a business case 33% of the time [1]. In addition to this, 27% of the time project sponsors admit that they exaggerate the benefits in order to get funding [2]. The implication is that at best, only 40% of projects have any clarity on the desired business outcome. The problem is that the business case is commonly treated as a hurdle to be jumped rather than as the first opportunity to seriously evaluate where effort should be directed. It is up to the board and the top management team to call out such behaviour and impose more rigorous discipline on investment decisions. Consider the following war story:

> **War Story – Lying to the Board**
>
> The board of a top-listed company was considering a proposal for an increase in the budget of an IT project that was fundamental to its international operations.
>
> The project commenced some two years beforehand. The task of developing a concept statement for the project was a part-time job for an executive with some business and IT experience. The project estimate was around $35M, with an NPV of net benefits of $50M.
>
> The next stage was the development of a feasibility study and business case. This stage was managed by a general manager with business and IT

Project Benefit Realisation and Project Management: The 6Q Governance Approach,
First Edition. Raymond C. Young and Vedran Zerjav.
© 2022 John Wiley & Sons Ltd. Published 2022 by John Wiley & Sons Ltd.

experience. This was his last assignment before retiring. At the end of this stage, the project team had grown to 30. The project cost estimate was $75M, and the NPV of net benefits had grown to $100M. It was close to a year since the concept paper was submitted – a lot of time had been wasted, and the benefit estimates turned out to be 'reverse-engineered' to meet required benefit targets.

The company's board approved continuing to a build stage, with an initial pilot site. The project was managed by someone with little direct project management experience. By the end of the pilot stage, the project estimate had increased to over $120M and the NPV of net benefits had increased to $200M on increasingly doubtful estimates. Another year had passed to get to this point. Over $50M had been spent ... though, surprisingly, no one could be sure how much had been spent ... the company did not have a project accounting system or a cost/schedule control system.

The project's benefits were in areas of new and increased revenue from overseas operations and reduced costs. It was fairly evident that the benefit estimates were way overstated and there was sufficient reason to believe that costs were understated. A 5-page paper was developed to brief the board and seek approval for a further increase in budget. The 5-page paper drew attention to the existence of a more detailed document that was available on request from the company secretary's office.

The 5-page paper was presented to a silent board by an executive director. The request was approved. One of the board members commented later that everyone knew that they were being lied to, but no one was prepared to ask the questions that would bring this issue out.

We have to sympathise with the board in the above war story. Asking Q1 is to open a can of worms. In the ideal world, the strategy gets determined at the beginning of the year and then it becomes the job of management to execute the strategy. Asking Q1 is to question the strategy and to ask Q1 once a project has started risks having to redo the strategy ... re-examining the PESTEL analysis, VRIN, 5-forces analysis and recalling the expensive consultants in their nice suits ... it could add months to the project.

The trouble is that at the beginning of any project, there are many unknowns. It is only as the project develops that the real issues become clear. The customers are unable to explain what they want in the beginning when the project is just an idea and other important stakeholders only realise the implications of various decisions as to the project proceeds. Think of Samsung's recent experience with its Galaxy Note 7 exploding in its customer's hands because its battery overheated.

There is almost always a need for planning and then some retrospective sense-making. We like to present our endeavours as meticulously planned, but the reality is often such that we adopt our plans to what is at hand. Therefore, there will have to be a continuous effort of comparing the original brief to the business strategy (that may have changed) and the project outputs that are being generated. It will be a dance between the emerging and the planned that will determine the best balance for what the project is about. Note however that the more complex the project gets, the less the planning will work, and the more change will have to be accommodated during the execution. This means that the project sponsor should favour serendipity and emergence a lot more in more complex and large-scale (both budget wise and time wise) projects and can stick to the original brief more firmly in fast-paced and more straightforward projects. This warrants very different leadership styles, the former being facilitative and enabling, the latter being more directive and goal-focused.

Often, however, there are ongoing projects for which the outcome is less than obvious. They will be burning resources for a long-established goal that may not even be relevant. Conversely, there is sometimes the situation that the world is very different from what it was when the project was sanctioned. COVID is just one example of this. Q1 focuses the mind to help us recognise these situations.

So, what do we do? In Chapter 3, we introduce AcdB, a tool we have discovered to help break down issues and help answer Q1? We like AcdB because it is quick to explain and can be done in an hour or so with the key stakeholders in the room. We introduce AcdB in Chapter 3 because there are many possible ways you can answer Q1 and we do not want to distract you, the reader, from understanding the need for the 6Q Governance questions. Q1 is necessary because strategy is often trying to solve a wicked problem and an iterative process needs to be taken to react as more information becomes available.

How to Know Whether Q1 has been Addressed Adequately

We finish the discussion of Q1 by reinforcing the message that although the relevance of Q1 goes without saying, you would be surprised how often organisations go amiss when addressing this question. We have seen all sorts of projects being stuck in their project 'delivery bubble' very quickly forgetting what the original intent of the project was briefly after getting the strategic go ahead. This doesn't mean blaming the project delivery team for focusing on the time or budget. It rather implies engaging in a continuous conversation about the project outputs and deliverables meeting the long-term strategic outcomes and making sure that the alignment between business objectives and the project deliverables is being met. Think of yourself as the guardian of the ever-changing business/policy brief. You might say that this is already translated into project objectives in the respective brief, but the reality is that policy and strategy change much more quickly than is needed for a project (especially large-scale and complex ones) to be delivered. Therefore, someone needs to engage with the project team to bring the strategic business perspective into consideration, to make sure alignment is there, and to make necessary adjustments at both ends of the strategy and delivery spectrum if the alignment is not there.

To illustrate this point, answer the following question to identify when Q1 has been addressed adequately:

Which of the following project proposals has clearly identified the strategic benefit(s)?

- Buy an exercise machine Y/N[1]
- Implement an ERP system Y/N[2]
- Build a bridge Y/N[3]
- Put a computer on every desk at school Y/N[4]

We have already pointed out that it is a fiduciary duty of the board and their delegates to ensure resources are spent wisely and that strategy is a responsibility of the board. In each of the examples above, the strategic conversation is not yet complete and as the sponsor, it is your job to call this out and insist that more thought is applied before allowing others to leap into action prematurely.

Q2. How Much Change?

The second 6Q Governance question, Q2, is to make an assessment on how much behavioural change is required to realise the desired benefits. The first 6Q Governance question Q1 is asked to seek clarity on what benefit is being targeted. The second 6Q Governance question Q2 is to evaluate whether the benefit can be realised or not. Recall that the unspoken truth is that half to two-thirds of projects either fail outright or deliver no discernible benefits [3]. Hence, the biggest risk of any project that the governors need to be aware of is that the benefits will not be realised and the strategy/policy is not implemented effectively. It is the responsibility of those charged with the governance of the project to focus on this overarching risk because those at the project management level need to focus on the on-time on-budget delivery of an output that is not the same thing as the realisation of benefits.

---

1 N, strategic benefit should be to lose weight/reduce heart rate etc.
2 N, strategic benefit should be to reduce costs/reduce stock outs/etc.
3 N, strategic benefit should be to reduce congestion/etc.
4 N, strategic benefit should be to increase literacy/improve PISA results/etc.

The particular insight of Q2 is to recognise that benefits are generally achieved through organisational and behavioural changes. It is relatively easier to deliver a new output, e.g. software, road, and bridge. It is much harder to get people to change to use the output in a way that the desired benefits are realised. It is a common error to define a project too narrowly around the delivery of output and forget about the need to promote behaviour change. KPMG conducted a study and found that change is considered only 40% of the time [4]. If their findings are more widely applicable, this is not appropriate. How often do you ask yourself at the time of funding a project 'what behaviour change is needed for this project to succeed?' If it is going to be too difficult to get people to change their behaviour, then there will be times that it is better to defer or even not to start a project. This is especially true when many projects are being undertaken concurrently because there is a limit to how much change people can handle. This last point is captured in the pithy saying: 'You can only eat four elephants a year'. If your organisation is already undertaking, say, three major changes, you need to be very sure of yourself before undertaking the fourth.

To illustrate how Q2 is applied in practice, we can consider an initiative taken by the Australian Government to reduce the impact of the Global Financial Crisis in 2008. The Government decided to throw money at the economy to ensure it did not stall and they directed a large amount of funds into the education sector. Every school was to have a new assembly hall whether they needed it or not, and it had to be built within a year of the funding announcement. If we apply the 6Q Governance framework, we might answer Q1 'what is the desired outcome?' with 'to stimulate the economy'. However, if we were an education department, this would be an unsatisfactory answer because our strategic goals include 'improvements in literacy and numeracy'. So, our answer to Q1 should be refined to something like 'to stimulate the economy AND improve literacy and numeracy'. With this clarity, we turn to Q2 and ask, 'How much change is required?' Another way of asking this question is 'IF we build a new assembly hall, will we stimulate the economy AND improve literacy and numeracy?' The answer is clearly 'no'. There is nothing wrong with building meeting halls to stimulate the economy, but meeting halls alone will not increase literacy and numeracy. Additional projects are required such as (i) training teachers on techniques to lift literacy or numeracy; (ii) scheduling sessions for poorly performing students; and (iii) building meeting halls that allow for such lessons to be delivered.

Q2 usually leads to the insight that a 'program' of projects is needed to achieve the strategic goal. In our example, we have to follow the direct instruction from the government to build meeting halls but we need to recognise and plan to fund the additional projects that are needed to achieve the strategic goal. As astute administrators, we might build the new meeting hall with movable walls so that additional classrooms can be configured when the meeting hall is not in use. Then we would stay on the alert for new opportunities to fund the other critical projects (a and b).

Again, for the sake of clarity, we have moved our description of the thinking tools that we have found useful to Section III. It is more important that you as the sponsor are looking for a programme of projects than for you to become an expert in change management and programme management tools. The tools themselves are unimportant; what matters is that there has been a thorough consideration of the behavioural change that is required to realise the benefits. Change rarely happens easily and your team needs to be clear of what change is necessary and what is planned to enable the changes to occur.

Q3. Sponsor

The third 6Q Governance question, Q3, is to ensure we have the right sponsor for the project. In many ways, Q3 is the most crucial governance question because the research has long recognised top management support (TMS) as a critical success factor for project success and, more recently, TMS has been found to be the most important critical success factor [5, 6]. Projects with sponsors committed to driving through the necessary change tend to be successful and those with uncommitted sponsors tend to be failures. Despite this finding, the project management literature almost completely neglects the need to find the most appropriate sponsor and focuses on the things a project manager can control.

Note, however, Q3 is not the first governance question because there are many examples of a sponsor driving through a project that is a pet project rather than a project of the highest strategic priority for an organisation. A strategy must come first.

Finding the right sponsor is a troubling question because as far as we know, there are no tools to help answer the question. The answer almost always comes down to judgement. The answer is also influenced by who is exercising their judgement:

1) The board or delegated funding body.
2) The current sponsor.
3) The project team.

We have decided in the absence of any tools, we will first describe the rules of thumb that should guide decision-making and then illustrate them through case studies.

Rules of Thumb

If you are:

1) *The board or delegated funding body:* assess how much the sponsor genuinely believes in a project and whether s/he has the influence and passion to

drive through the necessary change. Assessing the commitment of the sponsor is necessarily very subjective. The sponsor will generally appear very enthusiastic and committed at the time a project is presented for funding. The board, or its delegates, need to assess whether the sponsor is lying or deceiving himself/herself. Imagine, for example, it was possible to put a very large dipstick into the sponsor and then pull it out to find out how much 'bullshit' had been presented. Will the sponsor be there for the project when the going gets tough (and it always will)? The larger the scale of change required (Q2), the more influential and committed the sponsor needs to be. If the sponsor passes this subjective test, the project can proceed, but if it does not, then it is likely the project will suffer from a lack of top management support and eventually fail.

2) *The current sponsor:* ask yourself (i) Do you understand the change required if the benefits are to be realised?; (ii) Do you understand how to get change in the organisation?; (iii) Do you have enough influence to make the change happen?; and (iv) Whether you personally believe it will be worth the effort (knowing that there will be times the project will demand significant personal effort from you)?

3) *The project team:* assess whether the sponsor has the influence to make change happen and is personally committed to the project. There is nothing worse than finding out the sponsor is too busy when the going gets tough or that the sponsor is unwilling to use his or her personal power to try to influence key stakeholders to change.

A Special Warning

Not all managers will be well suited to the project sponsor role. In particular, managers that embrace the role of keeping things operating smoothly need to be particularly careful. In addition to this, many cultures reward people in roles that are adequate in relatively stable environments but are unsuitable for projects because of their fast pace of change. For example, Asian cultures place a high degree of value on preserving face and do not encourage an authority figure to be presented with new information that s/he does not know how to deal with. Other cultures are also guilty of this to varying degrees.

Projects by their very nature introduce change and uncertainty into a previously stable environment. A good project sponsor knows situations will arise which s/he may not have dealt with before and s/he will have to canvass the expertise of the project team and others to find a solution rather than being the sole authority on all things themselves. It requires a special kind of confidence and humility. (This discussion is continued in the section on Q5 on having the right project culture.)

We finish this discussion of Q3 by presenting a case of a project with exemplary project sponsorship. The case narrative is presented in normal text and the features to observe are in italics on the left in the box.

Case Study – SkyHigh

SkyHigh Property Investments is a subsidiary of a major investment bank. SkyHigh had trebled in size in only four years. The organisation had over 100 major properties, many thousands of tenants, and thousands of investors. The enormous growth in complexity was imposing operational stresses on the organisation. The CEO was acutely aware of it because two companies in the industry had recently lost market share because poor operational systems had undermined investor confidence.

The CEO was acutely aware of the strategic issue and was personally involved in recruiting two key people to address the problem.

Project Initiation

The CEO recruited a new chief operating officer (COO) and also recruited Paul Major from a competitor to work with the COO. Paul started by interviewing all 100 staff to identify the major issues. Paul found SkyHigh had 'bad systems, bad processes, bad support structure, and the wrong mix of staff'.

Senior management were not engaged at this initial stage, so the CEO authorised the sponsor to take the lead.

Establishing Project Governance Structure

A steering committee was established consisting of only the COO, Paul, the head of IT, and the head of finance.

A very small steering committee was established for rapid decision-making

Paul interviewed each of the key stakeholders to determine what the new system should do. Paul felt it was important not to restrict the list of requirements but made sure it was clarified whether a requirement was critical, nice to have, or a wish list item. In making this distinction, Paul made sure he understood the underlying business process.

To manage other commitments, each interview was restricted to about an hour and additional interviews were scheduled as necessary. In practice, it took about four interviews per stakeholder.

Very detailed background research of stakeholder requirements.

Case Study – SkyHigh   (Continued)

Once these initial interviews were finished, a 30-page document was prepared summarising all the requirements. Paul added his prioritisation of the requirements and sent it back to the stakeholders for confirmation.

Requirements were prioritised.

Paul said, 'I was very conscious that no package would be able to do everything and tried my best to manage expectations'.

Package Selection: Understand Workarounds and Trade-offs

The first key task was to compare potential replacement systems with the requirements document. 'We narrowed down the selection [quite quickly] to the two with the business clearly favouring one while IT preferred the other. We had to work through issue by issue before IT signed off on the final package'.

No one person has a complete understanding of the implications of various decisions, so it is necessary to have humility and acknowledge when your subordinates have relevant knowledge.

Paul was primarily a business user and recognised his weakness in matters of IT. He could have forced the issue, but he chose not to, recognising the important contribution IT had to make.

'[Neville] took the time to explain to me the implications of various decisions [in business language]'.

Strong communication skills are necessary in the search for a common understanding of the issues.

Neville said we had a standing joke at the beginning of each day 'Paul, time for your tutorial'. 'Paul wanted to understand and made the effort'.

As each issue was raised, Paul would systematically interrogate with the following set of questions: (i) Why is it essential?; (ii) What happens if it doesn't?; (iii) What happens if it breaches our standards?; and (iv) How can we work around this?.

The governance structure and the attitude of the decision makers clearly made a difference.

Sometimes the workaround was that senior management would formally acknowledge acceptance of certain risks, risks that the individual department could not accept. In these cases, Paul would talk to the COO (who would, in turn, talk to others as necessary), get approval in principle, and formalise this with an e-mail from the individual department noting the risk and a response from the COO accepting the risk.

The project sponsor and top management team must be willing to make time to understand risks and make decisions their subordinates cannot authorise.

(Continued)

Case Study – SkyHigh   (Continued)

In other cases, IT would develop the workaround solution even if it would cost them more time operationally. Their attitude was 'if this is what the business wants, and we can find a way to work around the problem, we shouldn't stand in the way even if it creates more work for us'.

In general, issues need to be resolved by good 2-way communication.
This issue highlights that the general approach followed was to discuss and try to convince. Interviewees commented:

'We always felt like we were heard ... there were lots of meetings and sometimes they went over time [to discuss issues that were important to us] ... We accepted the pressure to meet the deadline, but we were never pressured into accepting something we couldn't accept'.

However, there were also times a less conciliatory approach was followed. The COO admitted to saying: 'We don't want to hear this sort of argument' and it was reported that Paul would occasionally put a user in his place based on his knowledge of the overall business process and not allow someone to insist on inefficient current practice.

However, there are times when the sponsor and project manager need to cut short 'unenlightened' arguments
By the time the package was selected, there was a good understanding of what it could and could not do. The limitations were understood, workarounds had been developed, and the risks were acknowledged and accepted by the appropriate stakeholder.

The entire senior management team had attended a separate strategic briefing session with the potential vendors to evaluate their suitability as a long-term partner of the business. The successful vendor commented, 'I knew from the moment I walked in that this project would be successful' [because of the level of understanding and support from the senior management team].

The entire senior management team had a high level of understanding of the details of the project.
The commitment of the project team was very high, but the COO knew he was taking a gamble because 'we are going to change to suit the system'. All existing business processes were changing.

'You look at the culture, the willingness to change, and the perception of how big an issue you're fixing ... it took us a lot of time, a lot of communication, and a lot of discussions to reach this point... there was a desire to improve, we knew what we wanted from day one and we had everyone's buy in ... I knew the type of people I had and how far I could push them ... in a different organization, this would have taken twice as long [e.g. Government]'.

Case Study – SkyHigh   (Continued)

*You look at the culture, the willingness to change and the perception of how
big an issue you're fixing ... it takes a lot of time, a lot of communication, and a
lot of discussions to get everyone's buy in.*

Project Implementation: Monitoring and Managing Risks

A detailed project plan was developed, and the project was subsequently
monitored very closely against both the project plan and a risk management plan.

There was a lot of informal communication daily to ensure issues were
being raised at the regular weekly meetings of the project teams, and the
systems working group. Paul focused on one-on-one conversation, making
sure he knew what the real issues were while a project officer managed the
administrative processes.

*Although there is a formal project plan, project execution requires a lot of
informal communication on a daily basis to ensure issues are being raised at the
regular meetings.*

The steering committee met fortnightly on a formal basis mainly to ensure
the risks were being managed and to check that the benefits were still likely
to be realised.

*The focus at the steering committee level is on whether the expected benefits
are likely to be realised.*

Interviewees described this period as being almost incident-free albeit very
intense with very long hours being worked. They had their regular job that
included year-end reporting and also had to clean/correct their data from the
various existing systems and preparing it for upload into the new system.

Some in the finance department were also new to their jobs. No one
described any difficulties except for one incident. Paul recalls that at one
steering committee meeting, the COO noted to one of the other managers,
'you're falling behind', and the next day the problem was resolved. The COO
recalls saying something stronger in this incident: 'If you can't do it in time,
we'll find someone who can'.

*Although you need to work as a team, there are times the sponsor will need to
strongly exercise his/her authority.*

A detailed review of the minutes of the various meetings reveals that
unexpected incidents did occur with the potential to delay the project. The
implications on the overall project were noted, options for resolution were
discussed, and they were assigned to specific people for resolution. All the
incidents were resolved in two weeks or less. Interviewees attributed this to
the formal and informal governance structure allowing rapid identification

(Continued)

Case Study – SkyHigh   (Continued)

of issues, the rapid escalation of issues, the high level of the senior management ownership (willing to take decisions and accept risks) and the type of
the people on the project (driven to achieve).

In the final go/no go meeting, everyone signed off. It was made clear to
everyone that they did not have to sign and go live with the new system for
their module and that a workaround would be found.

> *The formal and informal governance needs to allow rapid identification of
> issues, rapid escalation of issues, high-level senior management ownership
> (willing to take decisions and accept risks).*

The COO said, 'The people who did this work are still with us now'. His
implication is that the way you manage this risk is by assigning tasks to
people who care and would be accountable.

> *Assign tasks to people who care about the result.*

Outcomes

A year after the implementation, all the interviewees consider the project a
success. Not long after the go live dates, the whole project team was taken
out to dinner to acknowledge their work.

In the project management terms, it is a clear success because it came in
on-time and below-budget and worked. 'A much higher level of data integrity.
The accounting close-off on 31 December was a far cleaner process. The
processes are maturing'.

> *The success of this project was strongly influenced by the quality of top
> management support at the CEO, CXO, and project sponsor level.*

Q4. Success Measures

The fourth 6Q governance question, Q4, is to determine how success will be measured. If Q3 has been addressed well, then the project will have a passionate and
influential sponsor. In many cases, this will be enough for success, but we believe
an additional governance mechanism is prudent – a formal measure of success
consistent with the project goal identified in Q1.

The reason why the board or their delegate needs to address Q4 is that the
default measure is time and cost. However, no project should be approved simply
to come in on-time and on-cost. This is a nice-to-have measure of the project management success and is the responsibility of the project manager. In contrast, a
measure of project success is the realisation of strategic benefits such as increased
profit, reduced cost, or improved customer satisfaction. In the public sector project, success could be measured by increased literacy, reduced crime, reduced traffic congestion, or reduced waiting time for health services. These project goals

should be the responsibility of the project sponsor and the operations staff who take over after the project team finishes.

There is a widespread misunderstanding that the project management success will lead to project success. This is untrue. There is only a weak correlation between project management success and project success [7]. For this reason, it is essential the board ensures there is someone focused on the targeted outcomes identified in Q1. If they do not do this, a force of habit will lead stakeholders to focus on time and cost and the targeted benefits will not be realised. According to some research, only 5–23% of the time is a sponsor made accountable for the benefits [8].

There is a second reason why appropriate KPIs are a board responsibility. The project sponsor needs to be formally made accountable for the benefits. If s/he is not made formally accountable, then the accountability will default to the project manager who will be measured on time, cost, and quality. No one will be formally accountable for the strategic benefits and they will not magically appear even if the project team does their job well.

There is also a political dimension to the KPIs. A politically astute project sponsor will know projects have high failure rates and therefore prefer not to be held responsible in case things go wrong. The sponsor may also be able to employ dysfunctional political responses should the project look like it is going to fail. For example, an astute project sponsor may arrange to be reassigned before a project fails. If this is not possible and if Q4 has not been effectively addressed, a sponsor may simply change the KPI to measure whatever has been achieved (and ignore whatever has not been achieved). In the worst instance, the project sponsor may simply shut down discussion of the project in the organisation and declare the project a success whether it was or not. For all these reasons, we recommend Q4 is addressed near the beginning of a project and the measures of project success be established as part of the overall project governance.

Rules of Thumb

There is extensive accounting literature on setting KPIs. However, as far as we are aware, there is no convincing evidence that one way is more effective than another. We feel it is effective enough to implement KPIs by following a few rules of thumb:

1) The KPIs must either directly or indirectly measure whether the targeted benefits are realised (Q1).
2) The KPIs must motivate the project sponsor to succeed.
3) The KPIs must motivate the project team and project stakeholders to succeed.

Some examples of the considerations that might be appropriate:

If, for example, the strategic benefit is to reduce labour cost – it would not be effective to set headcount as the KPI. Staff would be fearful of losing their jobs and would be unlikely to be motivated if success were to be measured by reduced headcount as the KPI.

Motivation is an extremely subtle 'science'. Paying a bonus if the project succeeds can only be expected to work in some environments, e.g. financial services. The evidence is quite strong that money should be considered a hygiene factor; it will demotivate if it is not enough, but it will not motivate once a certain threshold has been reached [9]. In a public sector environment, a more appropriate reward could be intrinsic such as more challenging work, or for the sponsor: a knighthood. Attention from a senior manager is very motivating (see Agency case study illustrating Q5 and Appendix C) which again points to the criticality of appointing the right project sponsor.

We finish this discussion of Q4 with a case study to illustrate the subtlety of having effective performance measures. The case implemented project performance measures at the board level, but they were treated as a hurdle to be jumped to secure funding and they were never taken seriously and managed through to realisation.

Case Study – TechMedia

TechMedia was established early last century as a semi-government entity to operate in a niche market of the media industry. It was quite a political organisation reflecting the entrenched culture, processes, and functional silos that had evolved over its long history.

[Project] TechMedia's culture was being changed through the appointment of a new CEO from the finance industry. Technology was one of the main tools underpinning the change. The year 2000 (Y2K) provided a convenient trigger to replace the financial system and further modernise the organisation. A steering committee of the organisation's most senior managers chaired by the CFO evaluated options ranging from a $250 000 workaround to a $10M Enterprise Resource Planning (ERP) system. They interviewed between 25 and 30 of the major vendors and found that although financial systems would solve the immediate Y2K issue, neither they nor the small ERP systems had the functionality to support TechMedia's operational activities. It seemed only the big ERP systems had the functionality to underpin their growth.

> *Even though IT was being implemented, the objective was behavioural change – to modernise the organisation.*

[Board reluctantly convinced] The preferred choice was presented to the board, but 'they were not convinced we could pull it off because they had all been bitten by an IT disaster in the past'. They initially needed to be convinced that buying a future product was a good strategy and then they kept asking for more figures and delayed making the decision for around 12 months.

Case Study – TechMedia   (Continued)

A senior manager felt strongly that the board should have decided earlier. 'They vacillated. They could have decided to upgrade [the existing package] earlier ... In this case, there was a lack of action. Decisions were almost made and then more justification was required. Because of this time lost, corners were cut, and benefits were lost'.

The board would have been more effective in their decision-making by applying 6Q Governance. The initial question, Q1, is answered based on strategy rather than time or cost.

Consultants were engaged to help justify the decision. They helped senior managers identify and individually sign off against $6 million of benefits to be realised over five years. The consultants reported 'these savings as conservative ... reflecting only 50% of the available savings' but added, '[TechMedia] will need to re-engineer their processes to take advantage of the opportunity offered in the technology selected'. The proposed budget of $10M was sold to the board partly on 'the mantra of survival' and partly on 'the huge benefits of ERP'.

The board 'was eventually forced into making a choice by Y2K' but the delay 'reduced the amount of time available to implement'.

Preparation of the business case was treated as a hurdle to be jumped. Q2 of 6Q Governance would have helped the board to realise a huge amount of change was required to realise the benefits.

[The Project – stated and unstated objectives] It was decided to implement in two stages. Stage 1 was to replace the financial system to meet Y2K compliance. Stage 2 was to follow and implement the other requirements. One manager described stage 2 as 'less important' but, in fact, would be where most of the benefits of the board justification document would be realised. The problem is that the general strategy of the organisation had been developed by relatively few people at the senior level. The project had been justified to the board for its business benefits, but it was justified to the organisation for its technical advantages. There was no consensus in the organisation of exactly what was needed and none of the interviewees explained the selection criteria in similar terms.

'We never really understood why the ERP was selected or what it was supposed to achieve'.

'The hot points of ERP were that we had 23 different systems, of which only a few talked. The sales talk was about replacement of 23 with one'.

The project had no clear answer to Q1 – what are we trying to achieve. The KPIs were financial at the board level and operational at the organisational level. Even then, there was no focus on the KPIs.

Q5. The Right Project Culture

The fifth project governance, Q5, is asked once the project commences. The question is whether you have the right project culture. With the right project culture, all the relevant information will be reported and decisions will be made on the basis of the knowledge being generated in the project.

The question is justified by the research literature which finds successful projects are not related very strongly to the quality of planning (as emphasised in traditional project management guidelines) but more to the ability of the project team to change the project plan as issues arise [10]. The logic behind this is quite simple; it is almost impossible to predict the future. Therefore, it is almost impossible to follow a project plan exactly as originally planned because that would require an anticipation of everything that could cause a deviation from the plan. It is more effective to do some project planning and then stay very alert to emergent events that may require the project plan to be modified to respond. The key is to stay alert to emergent events and respond appropriately. Thus, the project culture created by the project sponsor is enormously important. The discussion below continues the discussion started in Q3 and explores the project culture created by the project sponsor.

The project sponsor cannot be expected to be the first person to recognise an issue has arisen which has the potential to affect whether the strategic benefits will be realised. It is far more likely that a project team member or a project stakeholder will be the first to suspect something is an issue. However, a project team member or project stakeholder will rarely have the authority alone to act on an issue. In the first instance they may not even be sure it is a real issue so they will need to discuss it with other team members, and if they agree an issue is important to raise it for discussion in formal status meetings.

Note the quality of the issues raised is dependent on how well Q1 and Q4 have been answered. If strategic benefits have been clearly identified and communicated, then project stakeholders will be in a better position to identify which emergent issues are important and which less so. They will not always understand the strategic context well. So, some issues they raise will be unimportant, but the process is important. If the issues they raised are treated with respect, discussed adequately, and responded to in a manner that reinforces the importance of the targeted benefits, then further issues will be raised as they emerge. If the issues are not treated respectfully, or if it becomes apparent that time and cost are the only issues taken seriously, then the stakeholders are unlikely to raise further emergent issues. It is the responsibility of the project sponsor, project manager, and the entire team to shape the project culture so that all the relevant/emergent information is raised for discussion. Q5 can be paraphrased – 'are we getting all the relevant project information?'

The following case is presented again in discussion of Q5 and in Appendix C to illustrate the exceptional project culture created by a project sponsor and her project manager. It will follow the format introduced earlier with the case in the left column and the commentary in the right column. The key points of the case are to illustrate how uncertainty influences decision-making in projects and how teams need to adapt and change plans to respond to emergent conditions arising in the project. The case illustrates the crucial role of the project sponsor in creating a culture where stakeholders are not afraid to raise issues as soon as they become aware of them.

Case Study – The Agency

The Agency is a world-leading public sector organisation providing scientifically derived information services. Agency staff are constantly at the leading edge of technology (e.g. high-end supercomputer projects) but their focus has always been to improve operations rather than simply exploring the technology.

A world-leading organisation.

The government issued a directive for accrual accounting to be introduced. A second directive was to standardise computer systems across the whole of the government.

The government mandated a new financial system but there was a lack of skills in the finance department.

A new Finance Director was appointed. She was the first senior manager to be appointed from outside the Agency in many years and the only female senior manager. She reports that it took her some time to get to grips with the issue. She 'would not be panicked into making a decision' and needed to reach the point where 'there was real ownership' on her part. The problem was that there was 'a serious lack of skills' in the finance department (who did not understand accrual accounting) and there were conflicting and strongly held opinions for the best way forward.

The project sponsor gave herself time to develop real ownership of the project.

The decision was made more difficult when her peers would stop her in the corridor and ask her when she would make a decision about the financial system. She realised:

'I didn't get a lot of time with the Agency Head ... I knew I wouldn't be able to change them [and get their support] ... so I had to reach the point where I was convinced the project had to succeed and then ... I would do it with them or without them'.

(Continued)

Case Study – The Agency   (Continued)

Henry Tell, her senior financial officer, was of the strong opinion that the existing system should be upgraded. However, this option would result in a non-standard system and was risky because 'so much money had already been spent on the existing system without providing a satisfactory system'.

Conflicting information.

Henry had commissioned a consultant to explore the issues, but the computer support manager felt strongly that the main recommendation (to develop a new system in-house) was beyond the capacity of the Agency. Mark Black, the project manager of a 'failed' MIS project, commented in writing that the consultant's advice to develop a new system was more complicated than might first appear.

The first overt sign that the Finance Director was committing to the project came almost a year later when she formally requested funding from the MIS steering committee. The request was to undertake the first of the consultant's recommendations (to determine the data requirements and functionality needed) with the intention to replace/upgrade/outsource the Agency's financial systems.

The request to initiate the replacement system was a positive move, but it must also have been received with some angst at the senior management level. Funding had been set aside for the upgrade of the financial systems, but (i) senior management and the Agency Head, in particular, did not like to 'waste money' on financial systems; (ii) the track record in Finance Branch was poor; and (iii) there was no reason to expect the Agency would not be the next 'horror' story given the bad experience of their colleagues in other public sector agencies.

A senior manager became 'very important to her' – Brian Minister. Brian had served for several years as the Agency Head's right-hand man in respect of budget matters and is an organisation troubleshooter often assigned to the most difficult assignments. When Brian openly supported the project by joining the finance system steering committee, it sent a clear signal to the rest of the organisation.

Influential stakeholder joins steering committee.

The working group reported to the steering committee that the option to outsource or to develop a new application was not viable. The main reasons were that there was nothing to outsource and there was not enough time to develop a system in a house that would be capable of full accrual accounting reporting by the required date.

The steering committee accepted this recommendation and agreed that a mini-scoping exercise be undertaken to evaluate various commercial

Case Study – The Agency   (Continued)

off-the-shelf financial packages. The option of upgrading the existing system was also to be evaluated.

These options reflected a major issue that had been simmering in the background. Mark Black (the project manager of the technically successful but poorly perceived MIS project) and Henry Tell (the champion of the existing bug-ridden system) had very different and very strong opinions of what the final solution should be. Mark felt that the system should not only report financial data for external purposes but also provide information for internal purposes. Henry opposed any additional complexity and believed that the best option was to enhance the existing system. The issue was initially resolved when the steering committee supported Mark's view. Henry however did not agree with this decision and continued to actively promote his views. Some of the documented incidents suggest a high level of tension:

- Henry refused to participate in the formal evaluation of some packages despite direct instructions in writing to attend.
- A file note recorded a conversation where Henry accused a colleague of interfering and ended with verbal abuse.
- Henry lodged a formal accusation of impropriety (in the selection process for packages to be evaluated) against the Finance Director.

Soft skills needed to overcome dissension.

During this period, there was a great deal of corridor talk. There were undercurrents at one level suggesting the viability of the Agency was being threatened because the new system would not work and at another level by suggesting that they were being unpatriotic by not supporting a local developer. Senior managers were never confronted directly with these misgivings but they 'picked it up in the corridors'.

The Finance Director dealt with the undermining rumours by calling meetings to air issues 'not knowing what would come out'. In the case of the formal accusation, the Finance Director arranged for the two accusers to have the chance to present their grievances directly to the MIS steering committee and then for the Agency to seek an external audit to clear itself of the accusation.

This period ended when both Henry and another opposer took leave and, finally, left the Agency. The issue was managed sensitively to show respect to all parties and while difficult to handle at the time, became relatively unimportant in the grand scheme of things. However, the resolution of the issue delayed the project by 6 months.

The working group recommended that the Agency purchase SAP. The recommendation was logical, well justified, and documented but the

(Continued)

Case Study – The Agency   (Continued)

unspoken feeling in the MIS steering committee was 'that other government agencies had tried and failed with SAP ... why would we be any different?' The recommendation was even harder to accept because the expected budget was a relatively large sum for the under-funded Agency.

Are we going to be the next failure?

A Cautious Phased Approach – Phase 1: Proof of Concept

The MIS steering committee approved the project, but they directed a cautious route. It was very important for them to avoid being in the next disaster. The experience of the Agency is that projects go wrong when insufficient thought has gone into the preparation and they wanted to undertake (i) an initial proof-of-concept phase to give them the option of terminating the project before committing to (ii) a full purchase and implementation.

The proof of concept was relatively trouble-free because Mark had been intimately involved in the preparation of the functional specifications, he understood the business processes well and his years of experience within the Agency prepared him to anticipate future needs. He was also aware of the major difficulties other agencies experienced. The end of this phase produced the Blueprint, a conceptual map of how each business process could be configured in SAP.

Phase 2 – Implementation

A detailed project plan was prepared relying heavily on the advice of the external consultants. Unfortunately, the plan that was prepared did not reflect the organisation realities at the Agency.

The problem was that of the fifteen staff required by the plan, only seven staff were really available to make any significant contribution to the project. Mark knew better than to ask for the 'best' staff to be allocated to the project. He reports that Agency managers would incredulously ask 'you want more?' to his requests for the basic level of representation. Some project team members described themselves as a 'second eleven'. Mark defended a few colleagues, but he acknowledges 'he would have liked more senior people in the team'.

The understaffed project had almost no chance of following the consultants' project plan.

Creating an Environment to Succeed Against the Odds

The deadlines looked very shaky and the viability of the project was being challenged. Mark recalls how important senior manager support was to him at this time and their intimate knowledge of the organisation.

Case Study – The Agency (Continued)

Mark would discuss issues with Brian and the Finance Director as he needed. With the issue of staff shortages, they either (i) developed strategies to ask for alternative resources; or (ii) reduced the scope of the project in a way likely to be acceptable to the board.

Both Brian and the Finance Director offered significant levels of support by making themselves easily available to discuss issues, ratifying decisions or contributing to alternatives and helping to sell decisions. Mark particularly valued the strength of the Finance Director's commitment to the decisions made and she made sure Mark had the confidence to know that she would personally fight for them. They had by then developed a good working relationship, understood how the other thought, and 'could communicate on the same wavelength'.

The Finance Director had an office on a different floor of the building, but she made a special effort to visit the project team as often as possible to 'shine on them'. She worked at creating a positive environment where people could rise to the challenges. She continued to call meetings to air issues whenever she sensed unease. She said, 'as time went on, people started to want to come on board as they could see it starting to work'. She did such a good job she even had to manage 'some jealousy because [the perception was that] the project team got all the attention'.

The Outcome

When people were asked whether the project was a success, no one gave an unqualified yes or no answer. In Agency terms, nothing remarkable happened. The project did not fall over, it was live, and working from day one with no major dramas. It met the revised go-live date and was within budget. Mark said, 'I didn't expect anyone to thank me because in the Agency, this is normal'.

The Finance Director remembers however a very subtle indicator of appreciation. 'The Agency made a modest contribution to costs associated with the celebration lunch'. It was symbolically very meaningful because this almost never happens and to her, it indicated that the Agency recognised the achievement was out of the ordinary. One of the main concerns of the executive was to avoid a disaster and it is significant that every member of the project team has subsequently been promoted.

In terms of the external benefits, it is now complying with accrual-based reporting requirements in a timely manner and audits have been completely unqualified. There is a high degree of confidence in the information produced.

In terms of internal benefits, however, the weakness in management reporting even two years after the implementation mars the overall success.

(Continued)

Q6. Monitoring

The sixth project governance question Q6 is related to the first and fourth questions: are we on track to realising the benefits? Like Q5 (project culture), Q6 (monitoring) is asked once a project has commenced and the question is necessary because it is very unusual for a project to proceed exactly as planned. There needs to be a way to ensure projects react to emergent events in ways that ensure the target benefits are still realised.

The literature suggests only 13% of the time are projects tracked through to benefits realisation [1]. Similarly, in a preliminary survey conducted for this handbook, none of the board members interviewed stated they have an effective process to cancel failing projects. This supports our assertion that Q5 and Q6 are important for effective project governance.

However, organisations are very different in their maturity and it is necessary to take organisational context into account when designing the monitoring mechanisms. If an organisation is immature in terms of its governance, then it might be most effective to simply rely on Q5 and informal governance to monitor progress against the goal. We continue below with a case study to illustrate how NOT to informally monitor progress and introduce a tool that might improve the outcome. When an organisation is more mature in terms of its project portfolio practices, it is possible to introduce more formal monitoring mechanisms through a Project Management Office, Internal Audit, or through eternal consultants. In Section III, we will introduce a case study of the Australian Defence Science Technology Group (DST) to illustrate an industry-leading example of how best to use formal governance tools to monitor projects. Before we do that, the case of TechMedia is reintroduced below to illustrate how NOT to informally monitor progress.

Informal Governance

The TechMedia and SkyHigh cases illustrate the difference between passive and active monitoring. TechMedia management were not clear on what they were

Case Study – TechMedia (contd. from Q4)

TechMedia was implementing technology to modernise the organisation. A working group of TechMedia's top management had evaluated the options and were recommending a $10M ERP system to their board. The board were unconvinced because they had all experienced IT project failures in their past. To satisfy the board's questions, the working group appointed consultants to help justify the decision. They unwisely chose to emphasise $6M of benefits that could be realised and did not identify any strategic reason for the project. Time pressures forced a decision to purchase the ERP, but the staff at lower levels of the organisation did not understand what it was supposed to achieve. Many thought it was to replace a large number of disparate IT systems.

Q1 (strategy) and Q4 (KPI) were not answered well and there was no clear vision of what was to be achieved.

A senior manager was particularly difficult. He would always state that he was committed to the project, but he allowed the ERP to be implemented in his area with a number of shortcomings. Because of his seniority, the project team was unable to force issues with him. He disagreed with some of the early design decisions and felt that he was 'being shouted down by the other members of the steering committee'. He withdrew psychologically from the process and eventually left. His replacement could not understand why processes in his area had not been re-engineered and started to blame the system, which was significantly slower than expected. The CEO was briefed on the issue, but it did not become clear to the CEO that he had to intervene until it was too late.

Q5 – The CEO was made aware of an issue with one of his senior managers, but he did not intervene even though he was the only one who had the authority to resolve the issue.

User acceptance was delayed because the technical problem took almost nine months to resolve. However, the entrenched culture of TechMedia appears to have compounded problems. Two similar functions were performed in different functional groups but 'the groups didn't communicate to realise that they should have integrated ... and what happens now is that information has to be double entered'. Several major reengineering initiatives in one division were not going well and senior managers concluded that too much was being attempted at one time and some initiatives had to be deferred. The board accepted the management's assessment of the situation but were quite unhappy that the technical functionality had been implemented without the benefits being delivered.

Senior managers from different functional groups did not communicate and no one forced the issue.

(Continued)

Case Study – TechMedia   (Continued)

> *Eventually, the board had to accept that the promised benefits would not be realised.*
>
> Minutes of the steering committee noted two major risk items for the entire length of the project. Mitigating actions were never taken, both risks eventuated, and the predicted difficulties occurred.
>
> *Monitoring alone is not enough – no one acted on two major risks identified at the steering committee level.*

monitoring against (i.e. the benefits they wanted to realise) and did not recognise when top management intervention was required when risks emerged. In contrast, SkyHigh management were very clear on what they were trying to achieve and whenever issues arose that could not be resolved by the steering committee, they were escalated by the project sponsor and resolved between the relevant top managers. The formal monitoring mechanisms were complemented by the informal governance of the senior managers and the informal governance to resolve issues was as important as the formal monitoring to recognise risks.

It is important not to overburden a project with formal governance mechanisms if it does not fit naturally into the culture of an organisation. In many cases, it will be enough to have some standing items in a steering committee agenda. We refer the reader to Section III for some of the formal and informal tools that can be used for monitoring.

References

1 Ward, J., Taylor, P., and Bond, P. (1996). Evaluation and realisation of IS/IT benefits: an empirical study of current practice. *Eur. J. Inf. Syst.* 4: 214–225.
2 Lin, C., Pervan, G., and McDermid, D. (2005). IS/IT investment evaluation and benefits realization issues in Australia. *J. Res. Pract. Inf. Technol.* 37 (3): 235–251.
3 Young, R. (2006). What is the ROI for IT Project Governance? Establishing a benchmark. *2006 IT Governance International Conference*, Auckland, New Zealand (November 2006).
4 KPMG (2005). *Global IT Project Management Survey: How Committed Are You?* KPMG.
5 Young, R. and Poon, S. (2013). Top management support-almost always necessary and sometimes sufficient for success: findings from a fuzzy set analysis. *Int. J. Proj. Manag.* 31 (7): 943–957.

6 Young, R. and Jordan, E. (2008). Top management support: mantra or necessity? *Int. J. Proj. Manag.* 26 (7): 713–725.

7 Markus, M.L., Axline, S., Petrie, D., and Tanis, C. (2000). Learning from adopters' experiences with ERP: problems encountered and success achieved. *J Inf Technol* 15 (4): 245–265.

8 Lin, C. and Pervan, G.P. (2001). A review of IS/IT investment evaluation and benefits management issues, problems and processes. In: *Information Technology Evaluation Methods and Management* (ed. W. Van Grembergen), 2–24. IGI Global.

9 Pink, D.H. (2011). *Drive: The Surprising Truth About What Motivates Us*. Penguin.

10 Dvir, D. and Lechler, T. (2004). Plans are nothing, changing plans is everything: the impact of changes on project success. *Res. Policy* 33: 1–15.

3

Tools and Techniques

Q1. Strategy – Diagnostic Toolkit

We start this section by introducing the first of our tools for the diagnostic toolkit. This tool is to help link strategy formulation and implementation. Project managers do not have to be experts at strategy formulation, but they are part of strategy implementation and they need to be able to contribute to this discussion as issues emerge. This is especially true because strategy is usually responding to a wicked problem.

There are many strategy tools, possibly as many as there are strategy consultants, but the one we like the best is AcdB a design science tool to help facilitate discussions to clarify the outcome. The reason we like it is that it is simple to learn and does not get in the way of the more important strategic conversation.[1]

The conversation is important because strategy, business, and management are seldom like maths or science; there is no one right answer. The goal in topics such as strategic planning, investment decisions, cultural change programs, and design of new systems and processes is not to find the 'right' answer, but to create a compelling argument about which option is best out of a range of possibilities. The great insight that the Greeks had – and which we have forgotten in the modern world – is that the key tool for addressing this kind of problem is conversation. But not any conversation, a good conversation must be broad enough to allow the full scope of the topic to be explored but the conversation must also converge to reach a conclusion. It is an art rather than a science because conversations by their very nature are fluid and flexible. Yet there are definitely some natural rhythms and

---

1 The remainder of this section has been written or paraphrased from text written by Julian Jenkins, formerly from 2nd Road Consulting, and now CEO of The Upfront Thinking Company.

Project Benefit Realisation and Project Management: The 6Q Governance Approach,
First Edition. Raymond C. Young and Vedran Zerjav.

structures to a good conversation and a set of simple tools that can be learned to make conversations far more productive than they often are.[2]

- In the early stages of the thinking process, we are confronted with a confusing array of issues, questions, challenges, and opportunities. The organisational context is often complex, and the problem itself can have many layers. Moreover, different people have different perceptions of where the issues lie.
- In this environment, the sort of conceptual skills required are fundamentally non-linear and intuitive, as we seek to push and prod our way through the mass of information and insights to find a pathway that will lead us forward. In most cases, the scientific analysis will not be very useful here, except perhaps to provide some data to consider. Real intellectual work involves reflecting on the experience, forming opinions, and weighing up different viewpoints.
- In the next stage, a strategic decision is made, or a hypothesis emerges as the best way to move forward. A clear direction has been established and momentum starts to build towards action. The decision or hypothesis may require some further creative thinking to develop it fully, but there is now a sense of purpose and increasing clarity.
- Finally, and only at the end, is it possible to apply the analytical, linear, systematic type of thinking that we often associate with planning, as we put in place structures, timeframes, and budgets.

Understanding the rhythm of a strategic conversation helps protect those running the meeting from falling into two temptations – the first is to get disheartened when things seem complex and fluid early on, and to fail to push through to a point of clarity and decision making; the second is to rush too quickly to making a decision out of the desire to move directly into action.

To have a strategic conversation, a group needs to spend time in four key places of thought, four 'conversation spaces', which we have labelled A, B, C, and D for easy recall:

A) Where are we now? A conversation needs to begin with a thorough exploration of the area, structure, system, or process under review. This part of the conversation involves reflection on past experience, the discovery of new insights about what is really going on, as well as identifying specific problems that need to be addressed.

B) Where do we want to be? Understanding the present helps us recognise the challenges we face but does not create momentum for change. This is why the 'B' space is so important. The group needs to shift gear into a different mental

---

2 http://www.gospelconversations.com/wp-content/uploads/2013/12/Pursuing-the-Art-of-Strategic-Conversation-2.doc.

space, one of imagination and aspiration. We cannot generate any enthusiasm for change without a vision of how the future could be different, or a dream of what we would like to see in place. The tension between the present ('A') and the future ('B') creates the momentum for change and engages the desires of the individual members of the group.

C) How do we get there? Knowing where we want to get to is a great step forward, but the conversation will ultimately remain fruitless unless we conceive some clearly defined strategies for how to get there. This involves both invention (conceiving what we could do) and judgement (working out which options have the highest priority or would create the most leverage). This stage of decision making and direction-setting is vital to crest the wave and build the downstream momentum.

D) What steps do we need to take? Only at the end of the process do we start working on an action plan, by defining what needs to happen next to put our strategies in place, what the timeframes should be, who we will need to engage and what resources we will need.

The methodology provides a clear structure and direction not only for the leader, but for the whole group. The approach can be explained in less than five minutes at the beginning of a conversation, and then serves as a useful reminder to the group about where the conversation is up to – especially for participants who want to jump straight into solutions and action before the conceptual thinking has been done. The only physical technology needed is a whiteboard to capture the conversation as it jumps from A–B–C–D and back again. The normal layout is shown in Figure 3.1 below.

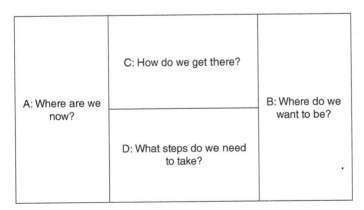

Figure 3.1 AcdB layout on a whiteboard.

Case: A 'Routine' Project Failure at TechServ

TechServ merged with Mainframe Pty Ltd. to form one of the largest computer services companies in Australia. The merger was a strategic response to the shrinking margins in the ICT hardware business, but it greatly increased the complexity of the business. To realise the benefits, ICT systems firstly had to be rationalised. Then it was crucial that staff costs were reduced by making process improvements and managed by matching staff levels against the profit margins in the various lines of business.

A complex integration of the first four legacy[3] systems had been achieved without any difficulties. Chris Little, the project manager, was to be promoted to a newly created CIO position as a reward. He believed his most important task for the organisation was to create a 'data warehouse'[4] to produce the information the company needed to manage its profitability.

Chris needed a holiday before assuming his new role. However, before taking his holiday, he was warned by two techies that another ICT systems integration project (of two other legacy systems) had a technical problem. Chris felt that the integration should have been relatively simple, but he checked with the project manager and the two senior managers responsible for the legacy systems. He was assured that it was all under control. He left to take his holiday but on his first day back, he faced a complete disaster.

A major functionality 'T&M invoicing' could not be processed in the newly integrated system and a second major functionality needed for customer satisfaction had been lost. During a crisis meeting, Chris considered reversing the integration but the project manager believed the problems could be fixed within a month. The Chief Operations Officer supported by the 2IC in the marketing both naïvely believed the project manager and argued against reversing the project. Chris was more interested in the 'data warehouse' and did not want to make unnecessary enemies, so he acquiesced. The discussion was all couched in technical terms and no one made the connection that T&M invoicing accounted for over 30% of the company's revenue.

Fixing the problems took many months longer than estimated. The company's image for customer service was permanently affected with an independent survey finding they had a bottom five rating many years later. The loss of 30% of its revenue caused the company to miss all its financial targets and all the senior managers lost their bonuses.

---

3 A legacy system is an old computer system that continues to work and need support. Legacy systems are often difficult to support because the people that made them aren't around anymore.

4 A data warehouse is a central computer repository that stores all (or significant portions of) the data collected by an organization's multiple business systems.

Case: A 'Routine' Project Failure at TechServ   (Continued)

In hindsight, it was apparent that the senior managers had overlooked many warning signs because of disinterest and naïve faith in the project manager. The COO's attitude was that he had little choice in the new system, so he was content to just leave it to the project manager. Another senior manager just did not care because he had lost the ICT function to the new CIO. The project manager was from the parent organisation and had no real stake in the outcome. She decided to 'just ram it in and fix the problems later'. The new CIO had misread the strategic priorities and was too weak to confront the project manager and his indifferent peers. All the senior managers were at fault for not trying harder to understand the business implications of the technical issues presented.

One of the perverse outcomes of the case was that the project manager was eventually promoted.

A case is now presented below to illustrate how the AcdB framework can be used to unravel the strategic issues surrounding a project.

Figure 3.2 shows how the AcdB heuristic was used to analyse the case of TechServ. In area A, 'where are we now?', the two issues which stand out are the fact that (i) there has been a merger of two IT service companies to create a more competitive organization; and (ii) there are too many legacy IT systems in the newly merged entity. In area C, the managers recognised they had to rationalise the number of IT systems and area D identified the traditional IT merger activities. The power of the analysis starts in area B where we ask: are we doing this? If

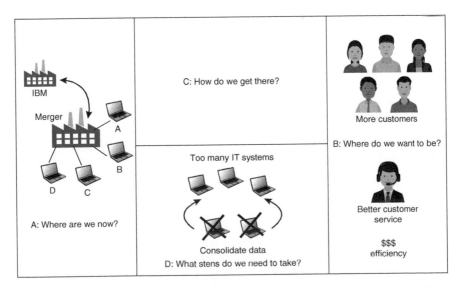

Figure 3.2   Worked example: TechServ! It's not an IT project, it's a customer service project.

the managers had taken the effort to think about it, they would have realised the objective was to provide better customer service across a wider range of IT products. It was not an IT project at all, it was a customer service project! With this insight, they might not have been so cavalier in charging ahead with the IT rationalisation and taken extra care every time a customer might have been impacted by their actions. It would have taken no more than an hour to perform an AcdB analysis, but it wasn't done and TechServ damaged its reputation in the market and undermined the whole reason for undertaking the merger in the first place. Strategy matters and it need not be hard to have a strategic decision.

Q2. Change – Tools and Techniques

There is no widely accepted way to assess Q2 'how much behavioural change is required to realise the desired benefits?' Change management is a very fractured discipline with many inconsistent messages [1] (summarised in Table 3.1).

Table 3.1   An overview of change management theories.

- Some, particularly those with a project management background, position change management as a way to minimise and control changes to the specifications
- Others understand organisation's need to change to adopt new innovations. However, the approaches to doing this are very diverse and confusing:
  - most build on the work of Lewin [2]
  - there is a general management stream which approaches change as just a part of what managers do [3–7], and this matured into concepts around the learning organisation [8–10]
  - there is a socio-technical stream [11–13] which explores how technical innovations are adopted [14] and matured particularly in Scandinavia into ideas around participative design and open systems theory [15]
  - there is an evolving discipline specifically called [organisational] change management which probably originated from Richard Beckhard and his book 'The Management of Change' [16] and was popularised by the major consultancies that inverted the title to change management
  - Roland Sullivan [17] states Organisational Development (OD) arose in response to the top down focus of Change Management to complement it with a bottom-up approach to embed change more permanently in organisations (and OD seems to be an extension of the earlier socio-technical systems theories but they do not seem to reference each other)
- While others still, probably at the extremes of most of the schools of thought listed, seem to position change as the ultimate objective and all other things as subservient, their approach is often derided as 'touchy feely' because the objectives can seem vague and the time span of change initiatives are conceived in terms of years and decades of effort.

Table 3.2   Examples to assess required change.

Output	Desired benefit	How much change is required?
Toll tunnel	Reduced congestion	Small/medium/large
Inner city cycle ways	Reduced congestion	Small/medium/large
New software (ERP)	Improved customer service	Small/medium/large
Company merger	Improved competitiveness	Small/medium/large

One effective way to overcome the lack of authoritative advice is simply to make an assessment based on experience rather than relying on the so-called experts. Table 3.2 provides some examples to test your ability to assess how much behavioural change is necessary.

Some readers will lack the experience to decide based on judgement alone or be uncomfortable with this approach and want a more rigorous method to guide decision making. Now that we have highlighted the need, new techniques will probably be developed to assess how much change is necessary but in the meantime, we can recommend what we believe is the best of what currently exists. The following list of tools is presented in the order of likely effectiveness (less effective to more effective), but readers should always bear in mind that they are only tools. Better tools may be developed in time and judgement will still be required to evaluate how much change is necessary to realise the strategic benefits.

- Stakeholder analysis
- Business process mapping
- Results Chain and Logic Model analyses
- Influencer analysis

Stakeholder Analysis

Stakeholder Analysis[5,6] is a well-known management technique and there are many excellent guidelines available to help prepare this analysis. This tool was not originally developed to answer the question of how much change, but it can be used to help answer this question.

---

5 https://en.wikipedia.org/wiki/Stakeholder_analysis#Methods_of_Stakeholder_Mapping. http://www.stakeholdermap.com/stakeholder-analysis.html. https://www.mindtools.com/pages/article/newPPM_07.htm.
6 https://en.wikipedia.org/wiki/Business_process_mapping.

Stakeholder analysis is a simple tool to identify the individuals or groups that are likely to affect or be affected by a project and to sort them according to their importance. Typically, the output from a stakeholder analysis is a power – interest matrix (Figure 3.3). The stakeholder analysis does not directly answer the question of how much change but it is helpful because it systematically identifies the key stakeholders and can help identify those stakeholders that have to change the most for the strategic benefits of a project to be realised. As a project sponsor, you will need to assess whether you and your supporters will have enough influence to get the key stakeholders to change. The most useful application of this type of analysis is to inform the membership of the project governance structures. Key stakeholders should be considered for membership of the steering committee; for example, the project sponsors of both the SkyHigh and Agency cases in Appendices A.2 and A.3 clearly considered who had the power and interest in their organisation and both opted for very small powerful steering committees.

Business Process Mapping

Business Process Mapping is the second tool we recommend to project sponsors and governance committees. The tool is commonly used for IT applications, but the tool is often too detailed for governance purposes. What is helpful is enough detail to compare the current process ('as-is') with the proposed future process ('to-be') as shown in Figure 3.4. The governing authorities can use this comparison to make an assessment of whether the proposed changes are small, modest, or large and, in turn, make an assessment of whether their organisation(s)

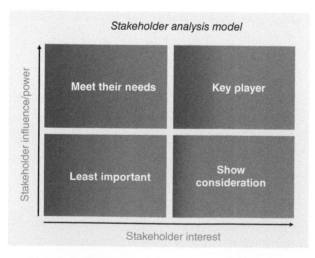

Figure 3.3 Typical output from a stakeholder analysis.

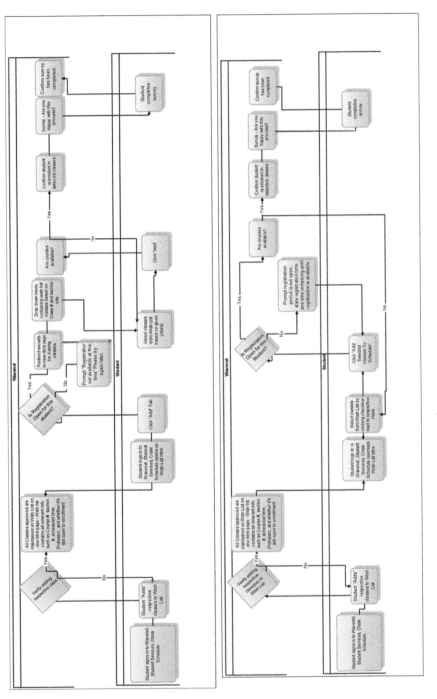

Figure 3.4 Helpful output from business process mapping.

will be able to implement the proposed change. Judgement of the capacity of the key stakeholders is still required but the advantage of this analysis is that it is based on a more systematic and detailed assessment of change than an assessment based on experience alone.

Results Chain or Logic Model

Results Chain or Logic Model analyses are a less common but more promising tool. The focus of these analyses is on the desired outcome rather than the output of a project. Both Results Chain and Logic Model are a way of describing and visualizing a strategy [18] and when done well, they identify what changes need to be delivered in addition to the outputs to realise the desired business benefits.

Logic models were originally developed for USAID in 1969 and their use was initially concentrated in the international aid arena. In 1996, the concept spread more widely to public policy, urban planning, and beyond. A Logic Model generally reads from left to right with Inputs feeding into activities, then outputs and then short-/medium-/long-term outcomes (Figure 3.5). For the logic map to be credible, it should have identified the necessary organisational and behavioural changes in the list of activities; and the outputs in the centre should logically lead to the desired outcomes on the right.

A Logic Map is sometimes referred to as a Results Chain, but there is another type of Results Chain that has its origins in the commercial sector [20]. This

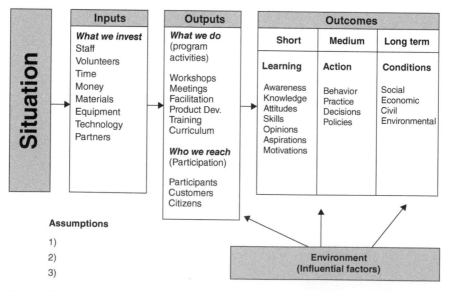

Figure 3.5 A typical logic map. *Source:* Arnold [19].

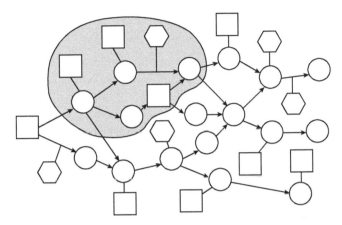

Figure 3.6   An example of a Results Chain. *Source:* Thorp, J. (2003).

second type of results chain is more detailed and uses three symbols (rectangle, oval, and hexagon) to represent Initiatives (i.e. projects), Outcomes and Assumptions, respectively (Figure 3.6).

This second type of Results Chain has the advantage of being more thorough and thought through than the typical Logic Map, but it has the disadvantage of being more difficult to read. To interpret the Results Chain, readers should understand the outcomes on the right as the desired long-term outcome, the outcomes in the middle and to the left of the page as short- and medium-term outcomes and the rectangles as the projects needed to deliver the outcomes. Some of the projects should be specifically designed to promote organisational and/or behavioural change. An exemplary organisation that uses Results Chains to good effect is the Australian Taxation Office. Again, as the project sponsor, your goal is to use the analysis to understand 'how much change is necessary?' i.e. how large is the program of work and then to ask the related question 'can we do it?'.

Influencer Analysis

Our final tool is called an 'Influencer analysis' [21]. It is a new tool that we predict will have high degrees of effectiveness in assessing change requirements. However, we need to temper our optimism because insufficient time has elapsed for this technique to be fully proven. The influencer analysis is based on the work of Albert Bandura [22, 23], the father of social learning theory and the most highly cited living psychologist.

The influencer analysis is based on research of people who were successful in getting behaviour change when others all failed. The researchers found that the key to behaviour change is first to study positive deviance: examples of change when no one else achieved the same result. The objective of this study is to identify the 'vital behaviour(s)' that leads to the desired outcome. Some examples:

- To overcome the COVID-19 pandemic, it is important to implement physical distancing, avoid crowded spaces, wear a mask, and wash hands frequently.
- To fight malaria, it is important to get rid of mosquitos that carry the disease.
- To overcome the guinea worm epidemic in Africa, the vital behaviour was found to be to filter drinking water with a fine mesh before consumption.
- To overcome the AIDS epidemic in Thailand, the vital behaviour is to use a condom when visiting a prostitute.
- To overcome the problem of recidivism, the vital behaviour is to remove the ex-con from his old neighbourhood and put them to work with a buddy to support them.

With any project, the starting point is to identify the vital behaviour that is most likely to lead to the desired outcome. Sometimes the vital behaviour is easy to identify and, at other times, detailed research is necessary. Then the key is to get people to perform the vital behaviour. This is seldom easy, and it turns out that the most successful change masters use four or more strategies concurrently.

Patterson et al. [21] have developed a framework of the six sources of influence to promote the vital behaviour (Table 3.3) and they make the observation that if four or more sources of influence are employed, then change is almost certain to succeed. They add, if less than four sources of influence are employed, then change is increasingly less likely.

We suggest that each project maps its various activities to the framework in Table 3.3 to see if four or more sources of influence have been employed. If not, the conclusion can be drawn that the behaviour change is unlikely to be realised and the project should not be funded without modification.

Table 3.3 Six sources of influence.

	Motivation	**Ability**
Personal	Experience the vital behaviour directly or vicariously through stories	Train people to be able to undertake the vital behavior
Social	Use peer pressure to acknowledge and reward people when the vital behaviour is performed	Ensure there is someone to help when people have difficulty performing the vital behaviour
Structural	Design meaningful rewards for performing the vital behaviour	Change the environment to remove physical and any other barriers to performing the vital behaviour

Source: Adapted from Patterson et al. [21].

The following case study is provided to illustrate how the change tools might be used in practice. The case is presented first and then two of the tools are used to illustrate how the case could have been managed to achieve a better result.

Case Study – The Agency

The Agency is a scientific organisation with a track record of success with IT. They helped build the third (or fourth) mainframe ever built in the early 1950s and they are recognised internationally for their scientific expertise.

The subject of this case started when the Federal government issued a mandate for all public sector entities to move from cash-based accounting to accrual-based accounting. No one in the Agency wanted to spend their limited resources on a financial system. They wanted to focus on the science in which they were world-class. There was also no confidence in the finance department. The finance department was largely neglected by the rest of the organisation and they had developed three underperforming IT systems over the years. The Agency took pride in the fact that once staff started working with them, they would never leave because they could not find a better organisation. However, in Finance, they never left because they only knew cash accounting, they couldn't get a job anywhere else.

The Agency knew they could not ignore the federal mandate. So, they hired a new senior manager (from Defence) to take on the job. However, she took her time to understand how the organisation worked. Over the course of a year, she slowly came to the personal conclusion that the Agency needed a proper accounting system and it was not a waste of resources. People kept stopping her in the corridor and asking her when she would start the project, but she did not allow herself to be pressured until she was personally committed to the project. She realised: 'I wouldn't get a lot of time with the Agency Head ... and the project was at the bottom of the totem pole (in the organisation)'.

Three things help turn the situation:

- [A confidante] She spoke to the board member that helped recruit her who reassured her 'you can do this'
- [An ally] A senior manager who joined her to form a steering committee (of two people) and signal to the organisation this project is serious
- [An insight] One of the underperforming systems was only perceived as a failure because of a lack of senior management support. She could trust the staff member who led the project and get a better result by being more hands-on in her sponsorship.

(Continued)

Case Study – The Agency   (Continued)

The project started typically by identifying the functionality the new system should have and then comparing that to the functionality of various systems. SAP was chosen and the first major issue arose because a manager associated with the existing underperforming system insisted that the choice should be to further upgrade the existing system. He firstly spread rumours that the sponsor was unpatriotic and later accused her of impropriety. The sponsor, whose office was on a different floor to the project team, would come down to visit frequently to find out how things were going. Whenever she sensed something was wrong, she would call a meeting to surface the issues not knowing what would come out. A colleague said, 'She was very brave (in her approach)'. When the formal accusation was made, she dealt with it calmly and transparently and eventually cleared the air and allowed the project to proceed. The accusing manager ended up taking stress leave and finally resigned.

The Project continued by seeking help from consultants to develop a project plan to implement SAP. This led to the second major issue because the project plan specified a certain resource level and the Agency did not have that many staff to spare. Only the left-over and newly appointed people were allocated to the project ... they called themselves the 'second eleven'. The consultants and the project manager were constantly at loggerheads because they were missing project milestones. However, the sponsor continued to visit the project team frequently to 'shine on them' and help them rise to the occasion. The project manager, sponsor, and steering committee eventually tapped into the Agency's depth of experience with IT systems and radically changed the project plan. They eliminated the 'time-wasting' sign-offs and cut back on testing because they had staff that could build a super-computer and knew IT systems well enough to know when a system was ready to go or not (NB: This approach could not be recommended in a normal organisation). There were other problems that arose but all of them were resolved through open channels of communication with the project sponsor, the steering committee as necessary, and the dedication of the project team.

There were multiple factors at play in the eventual success of this project but when you compare the success of this project with the perceived failure of an earlier project with the same project manager, the only real difference is the support of the sponsor and her senior management ally on the steering committee. The conclusion is that the passion of the sponsor makes the biggest difference between success and failure and it is, therefore, essential to find a sponsor that truly believes in the value of a project and who will personally intercede when issues arise.

How could this project have been governed better? In particular, how could Q2 – change have been addressed more effectively?

Firstly, one thing we must note is that the beginning of this project did not address Q1 very well. Using the AcdB heuristic, the strategic logic seemed to be C = accrual accounting, D = ERP system. There was no B that was of genuine interest to the organisation. This can be illustrated in the following way (Figure 3.7):

What would have been better is to explicitly acknowledge A = limited resources, and then to build a consensus for a B such as 'information to seek increased funding'. This would change the nature of the project to focus not on the IT system which is just a tool (and of no direct interest to senior management), but to focus instead on the strategic application of the IT system. Why do we want it, other than the government says we have to? Once we have found a strategic logic that engages the key stakeholders, we can move on to address Q2 – how much change is required?

Initially, our understanding of the program of work may be very simplistic (Figure 3.8). However, when we ask ourselves 'will the project (ERP) lead to increased funding?', we will realise the answer is 'no'. An IT system by itself is of minimal use without trained users. So, we realise Figure 3.9 needs to be enhanced by a project to train the users. This need for training was not initially recognised in the original Agency project. It was added later and was considered by many to be inadequate even though twice as much training was delivered than originally planned.

However, if we add an additional training project, we still have not done enough to provide information to seek funding. We need prototype reports to understand how accounting information can be used to seek additional funding. The Agency

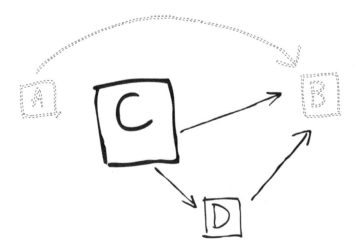

Figure 3.7   A weak AcdB logic.

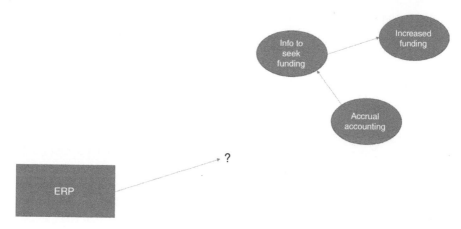

Figure 3.8   Initial value chain.

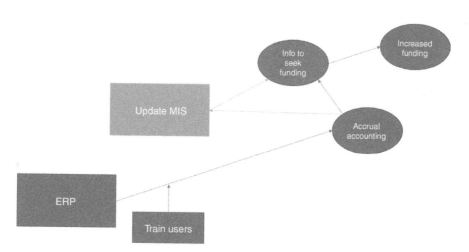

Figure 3.9   Enhanced value chain.

had a system they called the MIS which provided some management information. If an additional project were added to enhance the ERP system to provide something better than the MIS, then the Agency might achieve its goal – information to seek funding leading to increased funding. This enhanced value chain is shown below in Table 3.4.

From Figure 3.10, we can conduct an Influencer analysis to work out if we are doing enough to change behaviour. The vital behaviour is likely to be something

Table 3.4 Six sources of influence.

	Motivation	Ability
Personal	Form Tiger team to seek additional funding	Upgrade according staff knowledge. Train infrequent users. Develop cheat sheets.
Social		Establish a help desk.
Structural	Agency Head to reward/acknowledge Tiger Team if successful	Sponsor visits project team many times a week

Source: Adapted from Patterson et al. [21].

like 'use new system'.[7] In the ability column, we have 'trained users', but we may realise this is unlikely to be enough for a complex ERP system. The finance department staff may not have sufficient accounting knowledge to understand accrual accounting and they may need to upgrade their accounting knowledge. In the rest of the organisation, there will be infrequent users who will probably need a 'cheat sheet' to guide users on how to process the common transactions. The value chain can then be updated one more time to include the additional projects identified by the Influencer analysis (Figure 3.10).

What becomes apparent through this analysis is that it is rare that a strategic outcome (B) can be achieved by an individual project (represented by a single rectangle). It is far more common that a program consisting of multiple projects is required to achieve the strategic outcome(s). The governors need to make sure they understand the whole program of work before funding and they might be much better off not to fund any one individual project if they are not prepared to fund the whole program. Of course, there will probably be budget restrictions that mean the projects need to be funded over several financial periods, but in principle, if you are not prepared to fund the whole program, then it might be better not to fund only part of a program because if the strategic outcome is not going to be achieved, then what is the point?

It is worth mentioning another tool before we finish. The tool is called Program Management and at the first glance, it would seem to offer much promise because it is supposed to be focused on the realisation of benefits. However, our advice to you as a sponsor is not to put too much faith into the developed guidelines and to regard program management as simply another tool. Program Management has its roots in project management and the way it is practised can be too prescriptive

---

7 It is beyond the scope of this book to describe the Influencer analysis in more detail and readers are encouraged to refer to Patterson et al. [21] for more information.

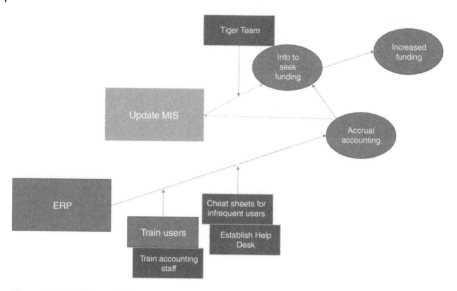

Figure 3.10   Value chain enhanced with projects identified through influencer analysis.

and document heavy. One study has even found 'mature' users of program management do not consistently follow its own methodologies [24] and we conclude the practice of program management is still in its infancy. It can be used to assess Q2 (how much change is required), but it should not be overemphasised and should not be followed religiously.

We conclude our discussion of Q2 with the observation that the question of how much change can be answered intuitively supported by analytical tools. Project governance is an art much more than a science and judgement is generally required. It would be a mistake to defer too much to the 'experts' and simply follow a methodology.

Q4. Measurement – Tools and Techniques

Ideally, the success measure should use data that is already collected or is easily collected. As we have mentioned in the earlier discussion, the key is to choose a success measure that motivates the right behaviour. The key is to invest your time in thinking: What motivates my key stakeholders? Which outcome of my project is likely to motivate my key stakeholders? How can I measure and report on this outcome efficiently?

If nothing appropriate is available, you may like to consider a change readiness measurement tool called ChangeTrack™ developed by a boutique consultancy acquired by Accenture [25]. This tool is implemented by surveying staff in an organisation to benchmark its progress against hundreds of other organisations that have undergone change and evaluating if the change is on track, going astray or headed for high performance.

Q6. Monitoring – Tools and Techniques

Table 3.5 presents a 6Q Governance™ tool that was developed for an organisation to use in its project steering committees. The tool was developed to help users assess in more detail which of the 6Q Governance™ questions were not being

Table 3.5   6Q governance diagnostic.

1) Benefits: 'keep the end defined'
1-2-3   Are we clear on our outcomes? (or have we lost sight of them?)
1-2-3   Is there a consistent understanding of the outcomes?
1-2-3   If our outcomes have evolved since initiation, has there been a formal acknowledgement of changes?
1-2-3   Are our project/program priorities consistent with our corporate objectives?
2) Change & Other Risks: 'they are all change projects'
1-2-3   Are we sure we understand how much change is required to realise the outcomes?
1-2-3   Have we planned for the entire program of work needed to realise our outcomes?
1-2-3   Have we got the support of all the key stakeholders?
1-2-3   Are we adequately addressing all the other major risks that you are aware of?
3) Sponsor: 'the key to success'
1-2-3   Does the Sponsor believe in the outcome and are they committed and passionate about driving through the change?
1-2-3   Does the Sponsor have the necessary authority/influence to drive the change?
1-2-3   Does the Sponsor have the competence/experience to drive the change?
4) Measures: 'motivate key stakeholders to focus on outcomes'
1-2-3   Are measures in place to evaluate whether the outcomes (vs outputs) are achieved?
1-2-3   Are the measures sufficiently motivating for the key stakeholders?
5) Project Culture: 'are there any warning signs?'
1-2-3   Is there a culture that encourages issues/risks to be raised?
1-2-3   Are decision-makers giving priority to issues/risks that may affect the outcomes?
(1 = lower priority issues (i.e. on time and on budget) get more attention and/or stated objectives are undermined)
6) Monitor: 'active or passive'
1-2-3   Are questions being asked to ensure the right information is being provided?
1-2-3   Is monitoring independent of the Sponsor and Project Team?
1-2-3   Are projects cancelled or re-scoped when outcomes cannot be realised?

addressed well in a project. Each 6Q Governance™ question was supplemented with a few diagnostic questions and users could assess using a traffic light system whether the diagnostic question was being addressed poorly – 1, adequately – 2, or well – 3. The tool should be used to identify which of the 6Q Governance™ questions were least well addressed with the understanding that it is unlikely that all questions would be well addressed at the earlier stages of a project. The tool is designed to help prioritise governance effort.

Formal Governance

In a more mature organisation, more formal frameworks may be implemented to monitor projects. Portfolio/program/project offices (P3O or PMO) are particularly useful if they are implemented to help governance and monitor benefits. However, in many cases, the PMO loses their way and becomes little more than template police focusing on process and the lesser issues of on-time and on-budget. It is important not to be a slave of the governance framework but rather to learn to focus on the targeted benefits (Q1), the reason why projects are initiated. The following case study of the Defence Science and Technology Group (DSTG) illustrates this and they are an example of what might be the world's best practice in benefits management.

Case Study – DSTG – World's Best Practice Benefits Monitoring

The Defence Science and Technology Group (DSTG) is the second largest research and development organisation in Australia with 2100 scientists and engineers on their staff. They are a world-class research organisation credited with leading edge developments such as the black-box-flight-recorder and the Jindalee-over-the-horizon-radar. However, two audits had found it difficult to transparently see the link between DSTG's inputs and outputs, and concluded that it needed to better articulate its value proposition and quantitatively demonstrate the extent to which its science and technology work aligns with the strategic priorities of Defence.

A major problem is that the demands for research far exceeded DSTG's capacity to deliver and prioritization of client-initiated requests for support tended to be from the bottom up. It was often driven by support requests defined by many individual stakeholders rather than by the strategic needs of multiple defence stakeholders.

DSTG responded to the audits by commissioning an independent review of its value proposition, which demonstrated that the top ten projects delivered over $5.1B of value to Defence; and by implementing a major project portfolio

---

Case Study – DSTG – World's Best Practice Benefits Monitoring   (Continued)

management initiative (PPM). Project funding was allocated (i) firstly to one of five major streams of investment, (ii) then to business units; and (iii) then to specific projects (Figure 3.11). All project funding was now to be justified annually by assessing the targeted benefits against the strategic objectives of DSTG and Defence as a whole. This was a major strategic initiative supported by senior management and implemented over a three-year timeframe.

The PPM initiative started with strategic stakeholder engagement. Multiple client stakeholders were invited to a workshop to vote on the relative priority of proposed DSTG support requests. However, this approach was of limited effectiveness because there were too many (about 1000) requests for stakeholders to assess. The PPM initiative then introduced a business case tool called an Investment Logic Map (ILM) to force DSTG scientists to define strategically related projects (no more than 20) and present them against program-level strategic defence goals (step 3 in Figure 3.11). The particular strength of the ILM is that it forced decision makers to identify the problem to be solved and the benefits to be realised rather than to focus exclusively on time and cost concerns.

Initially, this new investment approach adjusted current project budgets by no more than 10% to avoid major disruption while stakeholders gained familiarity with the new process. Training and change management was quite extensive and the introduction of ILMs was well received at the client level. Senior officers from different branches of the military could see the big picture of how the investment was proposed to be allocated. This allowed the discussion to be very open and trade-offs were made where officers would

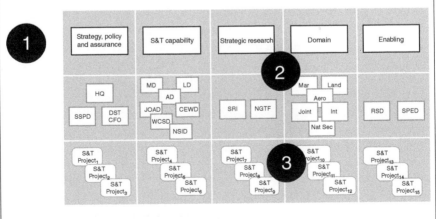

Figure 3.11   Prototype project reports.

(Continued)

Case Study – DSTG – World's Best Practice Benefits Monitoring   (Continued)

agree to funding and resource reductions in some areas in order to increase effort in other areas and target better overall outcomes. The second and third years of the PPM initiative were characterised by larger changes in the project budgets (modified zero-based budgeting) and historical resource allocations were no longer the deciding factor in allocating future budgets.

The project portfolio management initiative is now largely complete and DSTG is now trying to institutionalise their new investment processes. They are replacing their legacy IT system with an industry-standard project portfolio system (MS Project Server) and configuring it to capture the unique features of their new investment process. One aspect of their new system that may lead the world is the way they have implemented benefits management. They have configured their project management software to mirror the ILM and projects are assessed in the first instance on the basis of the problems to be solved and the benefits to be targeted. Then once a project is approved for funding the regular reporting for each project will always focus on whether the benefits are on track to being realised as well as the traditional aspects of time and budget.

First, projects are proposed and created in the IT system by creating a business case that follows the ILM structure. The screenshot below shows projects start by highlighting the problem they are meant to solve and the benefits to be targeted. Users then document the major project activities/phases and estimate high-level time, costs, and resource commitments. As the project proceeds, the system tracks progress against the original benefits and deliverables defined in the ILM. This is used to assure that the project focus and outcomes do not diverge from the agreed business case. Projects remain within the system after they are completed, and benefits may be realised and tracked within the system for years after project closure (Figure 3.12).

DSTG is progressing efforts to extend the reporting and tracking of projects as illustrated in the prototype report below. At the project level, the traditional measures of on-time and on-budget will be reported using a traffic light system. What differentiates the approach are the two new reports. Firstly, at the customer level, users will be surveyed after each output is delivered and the traffic light reporting shows whether the client is satisfied or not. Secondly, the prototype report, which may position DSTG as a world leader, is the report of benefits which assesses sometime after output is delivered, whether the expected benefits were realised. The expected benefits are specified in their project management system and benefits reporting is available at every stage of the project lifecycle. If DSTG projects are mainly green (on-time, on-budget) and customers are happy, but benefits are not being realised, then the

Case Study – DSTG – World's Best Practice Benefits Monitoring   (Continued)

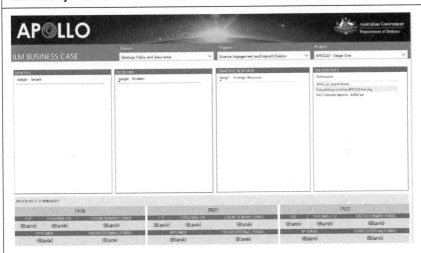

Figure 3.12   Screenshot of step 1 of the project initiation process – business case. *Source:* DSTG.

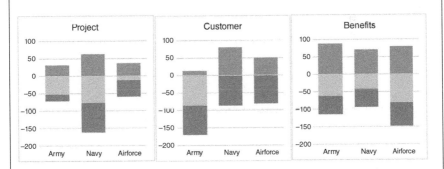

Figure 3.13   Prototype project portfolio management reports at DSTG.

leadership team will have a mechanism to identify this deficiency and intercede as appropriate (Figure 3.13).

The reporting infrastructure is only just being implemented as this case is going to print. So, it is too soon to say whether the design will be a complete success. Much will depend on senior management and what they choose to focus on in their regular project meetings. What we can say that if they end up focusing on whether the strategic benefits are being realised, then DSTG will have become an exemplar in benefits management as well in defence science research.

The value of the DSTG case study is that it shows effective monitoring is not dependent on religiously following a Portfolio Management framework nor having the best PPM software. DSTG implemented a small portfolio management office (PMO) and then focused on working with their top management to deliver their goals. In a different organisation, it might have been possible to achieve similar results by relying on Internal Audit or external consultants. Context is important and it is important to implement a formal or informal monitoring mechanism that fits the organisational context.

References

1 Cameron, E. and Green, M. (2015). *Making Sense of Change Management: A Complete Guide to the Models, Tools and Techniques of Organizational Change*, 4e. London: Kogan Page Ltd.

2 Burnes, B. (2004). Kurt Lewin and the planned approach to change: a re-appraisal. *J. Manag. Stud.* 41 (6): 977–1002.

3 Maslow, H. (1943). A theory of human motivation. *Psychol. Rev.* 50 (13): 370–396.

4 Wickstrom, G. and Bendix, T. (2000). The "Hawthorne effect" – what did the original Hawthorne studies actually show? *Scand. J. Work Environm. Health* 26 (4): 363–367.

5 Kotter, J.P. (2007). Leading change: why transformation efforts fail. *Harv. Bus. Rev.* 85 (1): 96–103.

6 Mintzberg, H. (1978). Patterns in strategy formation. *Manage. Sci.* 24 (9): 934–948.

7 Kanter, R.M. (1983). Change masters and the intricate architecture of corporate culture change. *Manag. Rev.* 72: 18–28.

8 McGregor, D. (1957). Human side of enterprise. *Manage. Rev.* 24: 41–49.

9 Senge, P.M. (2006). *The Fifth Discipline: the art and Practice of the Learning Organization*. New York: Doubleday/Currency.

10 Argyris, C. and Schön, D.A. (1978). Organizational learning: a theory of action perspective. *J. Appl. Behav. Sci.* 15 (4): 542–548.

11 Trist, E. (1981). The evolution of socio-technical systems. *Conf. Organ. Des. Perform.* 2: 1–67.

12 Trist, E. and Baumforth, K. (1951). Some social and psychological consequences of the Longwall method of coal getting. *Hum. Relat.* 4 (38): 7–9.

13 Mumford, E. (2006). The story of socio-technical design: reflections on its successes, failures and potential. *Inf. Syst. J.* 16 (4): 317–342.

14 Rogers, E.M. (1995). *Diffusion of Innovations*. New York: Free Press.

15 Emery, F.E. (1959). Characteristics of socio-technical systems. *Lond. Tavistock Inst. Hum. Relat.* 1959: 1–31.

16 Beckhard, R. (1969). *Organization Development: Strategies and Models*. Reading, MA: Addison-Wesley Publishing Company.

17 Sullivan, R.L., Rothwell, W.J., and Balasi, M.J.B. (2013). Organization development (OD) and change management (CM): whole system transformation. *Dev. Learn. Organ. An Int. J.* 27 (6): 18–23.

18 Moss, I.D. (2016). Defense of logic models. *Createquity* 2012 [Online]. http://createquity.com/2012/06/in-defense-of-logic-models/ (accessed 29 December 2016).

19 Arnold, M.E. (2002). Be 'logical' about program evaluation: begin with learning assessment. *J. Ext.* 40 (3): 3.

20 Thorp, J. (2003). *The Information Paradox: Realizing the Business Benefits of Information Technology, Revised Ed*. Toronto: McGraw-Hill Ryerson.

21 Patterson, K., Grenny, J., Maxfield, D. et al. (2008). *Influencer: The Power to Change Anything*. New York: McGraw-Hill.

22 Bandura, A. (1986). *Social Foundations of Thought and Action: A Social Cognitive Theory*, Prentice-Hall Series in Social Learning Theory. Englewood Cliffs, NJ: Prentice-Hall.

23 Bandura, A. and Walters, R.H. (1963). *Social Learning and Personality Development*. New York: Holt Rinehart and Winston.

24 Lycett, M., Rassau, A., and Danson, J. (2004). Programme management: a critical review. *Int. J. Proj. Manag.* 22 (4): 289–299.

25 Parry, W. (2015). *Big Change, Best Path: Successfully Managing Organizational Change with Wisdom, Analytics and Insight*. London: Kogan.

4

Further Insight

When Do You Ask Each 6Q Governance™ Question?

The 6Q Governance questions are logical and based on solid peer-reviewed research, but as the reader, you like we must have asked, will this really work? The 6Q Governance questions were derived inductively from five projects with a strong IT flavour. It seems reasonable to expect the 6Q Governance questions will be a useful guide for other IT projects but what about other types of projects such as engineering projects, infrastructure projects, marketing projects, organisational change projects, or mergers and acquisitions? Are five projects enough to guide the many tens or hundreds of other projects that we will undertake in our careers?

In 2019, Young *et al.* had the opportunity to answer this question by analysing whether the 6Q Governance questions correlated with project success in a large database of projects. This database was collected over a nine-year period and consisted of over one hundred thousand survey results from staff undertaking large change projects in 150 multinational companies. In terms of project type, one-third were concerned with IT implementations, and the remainder are organisational change-related projects. The complete range of organisation change types included: restructuring, business transformation, call centre, cost management, cultural change, general performance improvement, health services, human capital strategy, IT Transformation, leadership alignment, merger/acquisition, a new organisation, structures, outsourcing, process improvement, shared services, and new strategic direction.

The analysis mapped the survey questions in the database against the 6Q Governance questions and analysed whether the 6Q Governance questions correlated with the overall success of the project in terms of increased profits, reduced costs, or increased customer satisfaction. They also analysed when in the project lifecycle, the 6Q Governance questions most correlated with success (initial, early, middle, or late stage of a project). Their results are reproduced in Table 4.1 and discussed below.

Project Benefit Realisation and Project Management: The 6Q Governance Approach, First Edition. Raymond C. Young and Vedran Zerjav.
© 2022 John Wiley & Sons Ltd. Published 2022 by John Wiley & Sons Ltd.

Table 4.1 Correlation of 6Q Governance constructs with project success.

| 6Q Governance Question | Stage of project | | | | |
	Initial	Early	Middle	Late	Total
Q1 Vision	0.275^a			0.207^b	
Q2 Change	0.285^a	0.451^b	0.311^a		0.371^c
Q3 Sponsor	0.333^b				
Q4 Success measure				0.264^a	
Q5 Project culture					0.522^c
Q6 Monitor			0.671^c	0.507^c	

The maths is not complicated but is probably unfamiliar to the average reader. The key detail are the letters a–c in each cell which indicate how strong the correlation is with project success. The letters give an indication of the size of the relationship with project success.
[a] Significant at 10% level.
[b] Significant at 5% level.
[c] Significant at a 1% level.

Initial Stage of a Project

Q1, Q2, and Q3 were found to be the most significant project governance mechanisms at the *initial* stage of the project. The implication is that it is important to appoint a sponsor that will drive organisational understanding to the point of agreement with a vision and gaining acceptance of the need for change. Note that 'agreement with the project vision' is more important than 'understanding the project vision', a finding quite different to that advocated by the project management and change management literatures [1]. This is an important finding because it contrasts with the existing literature which emphasises project methodologies and tools rather than the trust and competence of leaders at all levels. Project success in the *initial* stage requires sense-making rather than project planning or communicating the vision (for understanding).

Initial–Early–Middle Stages of a Project

Q2 is an important project governance mechanism for success from *the initial stage through to the middle stage* of the project. Specific factors are managing change, trust in business unit leaders, and responding quickly to change issues. This research provides quite a clear picture that 'they are all change projects' and they need to be governed accordingly.

Middle–Late Stages of a Project

Q1, Q4, and Q6 are project governance mechanisms that correlate strongly with success in the *middle–late* stages of a project. It seems that as a project nears its completion, it is important to measure and monitor progress against the vision of what is to be achieved. These control mechanisms are not only useful tools to curb potential opportunistic behaviour but also valuable mechanisms to keep stakeholders informed about the project and able to react to changes in a timely fashion.

General Mechanisms Throughout the Project Lifecycle

Q2 change and Q5 project culture were found to be important across the entire project lifecycle. Change has already been noted earlier and we repeat the finding that 'they are all change projects'. Project culture, however, was found to correlate quite significantly to success across the project lifecycle as a whole but not at any specific stage in the project lifecycle. The implication is that there must always be alertness to unexpected information that could compromise the realisation of the desired benefits.

The Best Guidance Available

The evidence in this study ranges from moderately strong to very strong in its support for the 6Q Governance questions. Every 6Q Governance question is correlated against project success. The evidence for Q4 success measures is slightly weaker, the evidence for the other questions is strong to very strong. The particular insight of this study is that it suggests when during a project, each 6Q Governance is most likely to have an impact on overall project success.

More research is needed to be absolutely sure the 6Q Governance questions will lead projects to success, but we are not aware of any other guidelines that have been tested so rigorously. We suggest the work is sufficiently strong for the industry to adopt the guidelines. It may be that further research will find we only need five questions or maybe it is better with seven, but for the moment, what we are presenting is the best guidance available.

Reference

1 Parry, W. (2015). *Big Change, Best Path: Successfully Managing Organizational Change with Wisdom, Analytics and Insight*. London: Kogan.

5

The Future of Project Management and Governance

Where Do We Go from Here?

The way forward in many ways is very simple, but it requires that we recognise there are deficiencies in current practice. This will not be easy for our so-called 'experts' and it will not be easy for boards and top managers because we rely on expert advice. A major shift in emphasis is required:

Boards and top managers may have to accept that they personally have the most influence on whether a project succeeds or fails [1].

It takes courage to overrule the advice of our experts. We think it is useful in concluding, to remind ourselves of the argument we put forward at the beginning of this handbook: project management success is less important than project success. If we understand the history of why we are in the current situation, we can remain respectful of the advice we are receiving and also know when to exercise our own judgement to govern our projects effectively to implement policy/strategy and create value for our organisations. Our final section therefore covers the history of project management.

The History and the Future of Project Management

Project Management as a toolbox emerged in the 1960s in the construction, pharmaceuticals, and aerospace sectors. Those sectors are all heavily based on large-scale and difficult projects that require meticulous planning in order to be delivered within their expectations. As a result, a comprehensive set of tools and methods for planning and execution of projects was developed – which we now refer to as Project Management (PM).

Project Benefit Realisation and Project Management: The 6Q Governance Approach,
First Edition. Raymond C. Young and Vedran Zerjav.
© 2022 John Wiley & Sons Ltd. Published 2022 by John Wiley & Sons Ltd.

The basis of PM is to break down the overall project into manageable tasks with time dependencies and budget allocations [2]. The goal of PM is therefore to overcome challenges of project delivery through centralised planning and decentralised execution [3]. The result of the culture of detailed planning in PM is that we often see Gantt charts of thousands upon thousands of activities – wallpapered around the construction manager's office.

Although having such detailed plans provides discipline by defining clear objectives and accountability for tasks, they can also become a bottleneck for the execution – because updating a detailed plan can take a long time. The other fact of the matter is that project requirements are changing all the time – which makes it very difficult, if not entirely impossible to stay entirely true to the original plans. But projects are traditionally evaluated by comparing the execution with the project brief, so PMs are incentivised to keep avoiding changes in the requirements and stick to the project baseline, even when introducing changes is important for the alignment of the project with the client's strategic objectives. As a consequence, the mindset of "trying to make the world fit the project brief' has been more of a rule than the exception in the practice of the traditional PM paradigm.

The world that we live in is changing – it always has – but it becomes clear that a business cannot maintain its competitive advantage if it thinks of the project baseline as something that is set in stone and shouldn't ever change. The result is that there are plenty of reported examples of projects that monumentally failed to deliver benefits to their users and created white elephants – assets that are of very little use but expensive to maintain. All this calls for a rethinking of what PM is really for and what the 'next generation' of PM thinking looks like.

Some evidence suggests a shift from top-down and centralised systems of long-term planning to self-organising and agile networks for short term planning, in which decisions can be made both bottom-up as well as top-down. ICT is a good example of this where we are observing wide acceptance of agile software development methodology, as opposed to the traditional waterfall models. The latter being a sequential process, whereas the former being an integrated process where distinct activities take place at the same time.

Table 5.1 summarises this overview.

Conclusion

We have presented in this handbook a simple framework to allow Boards, senior managers, and their project teams to make the transition from current practice to a new arrangement where every stakeholder is able to sense and respond in a coordinated way to events as they emerge. The world is changing rapidly around us. We need a framework like 6Q Governance™ to help multiple stakeholders to

Table 5.1 The future of project management.

Now	New	Next
Project teams: Project teams are assembled based on expertise in technical disciplines with very little – if any – communication and coordination across the disciplines.	Instead of independent work packages for decentralised execution, we are starting to see some interdisciplinary engagement with key individuals who perform boundary spanning and knowledge brokering roles across the disciplines.	The step beyond boundary spanning is a new kind of expertise, which is no longer rooted in traditional disciplinary paradigms (engineering, architecture, construction, etc.) but a new breed of a project practitioner who is able to switch between the disciplinary mindsets.
Organisational structures: Planning is undertaken in a centralised fashion and execution is decentralised and decoupled. Competitive procurement systems are based on arms-length contracting.	There is a shift to relational contracting, where social and psychological factors (i.e. trust) are getting more prominence compared to the traditional legal factors. Such are the concepts of alliancing, partnering and Integrated Project Delivery	Relational contracting becomes the norm leading to an integrated project execution organisation. These organisations are eco-systems of businesses and other stakeholders who collaborate based on mutual trust and adding value to the project.
Leadership and governance: Governance is centralised, decisions are made by senior project leaders and delegated down the supply chain for execution.	A move towards decentralised governance, made possible by repositories of shared data, that allows project teams to make decisions for themselves in a network – rather than strictly hierarchical setting.	Governance decentralises even more and senior project leadership assumes the role of facilitating project teams to make decisions, which drive the project towards its execution.

understand what the priority is at any given moment. This over-arching framework guides each stakeholder to know when to make their contribution to realise the strategic objectives and to know when to be silent because an issue is less important than they previously thought.

It takes practice to recognise the important issues amongst the flood of information in a typical project. The appendices have been provided to present in full the cases used in this handbook. All the detailed project information is provided from the perspective of multiple stakeholders and you are invited to read through each detailed case and see if you can identify which issues are important and which issues can safely be delegated. Answers have not been provided but you are encouraged to post your thoughts and join the discussion at www.6qgovernance.com.

We are not so naïve as to believe that once you read this handbook you will instantly be an expert. We believe expertise will come with time and reflection and we have deliberately established www.6qgovernance.com to support your journey to success. Our goal is to establish a community of practice to allow you to get help as you need it and to provide opportunities for mentors in the board and senior management community to be recognised and for them to share their experience. We also intend to monitor the website very closely to support you in your governance journey and bring about a change in best practice.

Projects are important and becoming more so. Climate change is a pressing issue and in the short term, huge sums will be spent to overcome the effect of COVID-19 on our economies. The success of these projects will underpin our long-term prosperity. We believe 6Q Governance is a valuable contribution to this mission and we invite you to join us to use and refine the framework and solve these societal issues. Let us all move beyond secondary project concerns such as on-time and on-budget and learn to focus on outcomes that will make a difference to our businesses, our society and our lives.

References

1 Young, R. and Jordan, E. (2008). Top management support: mantra or necessity? *Int. J. Proj. Manag.* 26 (7): 713–725.

2 Morris, P.W.G. (2013). *Reconstructing Project Management*. Wiley.

3 Morgan, M., Malek, W.A., and Levitt, R.E. (2008). *Executing Your Strategy*. Harvard Business School Press.

Appendix A

TechMedia[1]

How Projects Fail 'Successfully'

Background

TechMedia was established early last century as a semi-government entity to operate in a niche market of the media industry. It was quite a political organisation reflecting the entrenched culture, processes, and functional silos that had evolved over its long history.

> A time traveller from the 1920s would understand [our] process ... every meaningless clause has its origin in something that went wrong in the past.

TechMedia's culture was being changed through the appointment of a new CEO from the finance industry. It had been led to commercialisation in the mid-1990s, incorporation by the late 1990s, and, finally, to a listing on the stock exchange.

Technology was one of the main tools underpinning the change. In 1996, it was an almost computer-illiterate organisation. By 2001, it had aggressively used ICT to help grow its revenue by 20% and reduce staff by 10%.

> In 1996, we were under-invested in IT. We were using felt pens & overhead projectors, we had no e-mail, only a few people had voicemail, and none of our professional staff had PCs. All memos were handwritten and passed to a whole floor of PC operators for typing.

---

1 https://www.researchonline.mq.edu.au/vital/access/services/Download/mq:10164/SOURCE2?view=true.

Project Initiation

The year 2000 (Y2K) provided a convenient trigger to replace the financial system and further modernise the organisation. The existing financial system had been neglected and a CIO was appointed to help deal with the issue.

> We had [our old system] from 1990–91. We didn't keep it up to date with upgrades ... by 1997, when Y2K was first talked about, we realised that [it] would need too much work to be Y2K-compliant.

The CEO was focused on the higher-level objective of listing on the stock exchange but he recognised Y2K compliance was an important intermediate step. The project had to succeed so that he could ask the board for more funding to seek listing. He felt that he was too far away from the operations to be heavily involved in choosing a new system but he knew that ownership and commitment from the business were crucial. A steering committee of the organisation's most senior managers chaired by the CFO evaluated options ranging from $250 000 a work-around to a $10M Enterprise Resource Planning (ERP) system. They interviewed between 25 and 30 of the major vendors and found that although financial systems would solve the immediate Y2K issue, neither they nor the small ERP systems had the functionality to support TechMedia's operational activities. It seemed only the big ERP systems had the functionality to underpin their growth, e.g. project management, sales/call centre and revenue collection.

Package Selection

The board agreed with this assessment and various members of the steering committee travelled to sites around the world to investigate different implementations. The ICT people had originally favoured one ERP system while the business people favoured another. It was only after a significant drop in price and a visit to the product strategists in the head office was a consensus reached on the second ERP.

The steering committee was attracted to the second ERP package because 'it could do everything you could ever want it to do'. They believed the process management features in the next release and the strategic direction of the product would better support the nature of TechMedia's business. They recommended implementing the current release as an intermediate step to meet Y2K requirements.

Board Reluctantly Convinced

The preferred choice was presented to the board, but 'they were not convinced we could pull it off because they had all been bitten by an IT disaster in the past'. They initially needed to be convinced that buying a future product was a good strategy and then they kept asking for more figures and delayed making the decision for around 12 months.

A senior manager felt strongly that the board should have decided earlier.

> They vacillated. They could have decided to upgrade [the existing package] earlier ... In this case, there was a lack of action. Decisions were almost made and then more justification was required. Because of this time lost, corners were cut and benefits were lost.

Consultants were engaged to help justify the decision. They helped senior managers identify and individually sign off against $6M of benefits to be realised over five years. The consultants reported 'these savings as conservative ... reflecting only 50% of the available savings' but added, '[TechMedia] will need to re-engineer their processes to take advantage of the opportunity offered in the technology selected'. The proposed budget of $10M was sold to the board partly on 'the mantra of survival' and partly on 'the huge benefits of ERP'.

The board 'was eventually forced into making a choice by Y2K' but the delay 'reduced the amount of time available to implement'.

The Project – Stated and Unstated Objectives

Within the steering committee, the CFO had never been a particularly active sponsor and the CIO had led much of the evaluation activity. It had become apparent that the project would affect much more than finance and the role of the project sponsor passed to the COO.

It was decided to implement in two stages. Stage 1 was to replace the financial system to meet Y2K compliance. Stage 2 was to follow and implement the other requirements.

One manager described stage 2 as 'less important' but, in fact, would be where most of the benefits of the board justification document would be realised. The problem is that the general strategy of the organisation had been developed by relatively few people at the senior level. The project had been justified to the board for its business benefits, but it was justified to the organisation for its technical advantages. There was no consensus in the organisation of exactly what was needed and none of the interviewees explained the selection criteria in similar terms.

> We never really understood why the ERP was selected or what it was supposed to achieve.

> The hot points of ERP were that we had 23 different systems, of which only a few talked. The sales talk was about replacement of 23 with one.

Organisational commentators said, 'it was a snow job from the IT people... they wanted the ERP on their CV'. Another said they were 'wined and dined to assist decision-making'. One of the interviewees added '[it might have been better] to spend it on a CRM, a document management system, or a web-based intra/extranet system. They are more important to a knowledge business like ours'.

'Business was looking for an ERP to show the rest of the world we are big enough to get an ERP ... it was a symbol of company maturity and size and signalled our coming of age'.

The CEO acknowledged the financial management functions ERP were not 'fundamental to the organisation' but said, 'even so ... I would do it all again'. The feeling expressed by many at the senior manager level is that '[despite the problems] it was a good investment that has served the organisation well'.

A False Start ...

An aggressive nine-month deadline was set for the implementation of Stage 1. A large project team of 34 staff and consultants was proposed. The consultants stressed that selection of key staff and empowering the project manager with resources and direct access to the CEO was critical for success. However, the contract was not signed. The consultants were unwilling to sign TechMedia's standard contract where the contractor would assume some risk for not meeting Y2K compliance.

When the contract was re-tendered to two other consulting organisations, the winning bid emphasised the use of their methodology to realise benefits. However, they could not commit to three of the originally promised benefits. The board accepted this change because the promised benefits were still more than that expected from simply upgrading the old system.

Stage 1

Three months were lost re-tendering the project but TechMedia was not able to extend the Stage 1 deadline beyond the original 1 July target because of limitations in the existing system. There was no choice but to implement the accounting module in only six months. Detailed performance bonuses were put in place for on-time implementation. The bonuses included senior managers through to the project team and including the consultants.

The project manager was reported to be very effective.

'The project manager was very strong' and 'would cause personality conflicts and not care in order to get the job done'. 'He pushed politics to the limits ... arranging to report directly to the CEO instead of to the CIO'.

> The project team was chosen internally because of their expertise. Incentives were in place, and we knew it had to be out by go-live ... They were strong, passionate and worked many late nights ... as a team for Stage 1, we worked really well.

Many project team members also acknowledged the value of the consulting input. 'In addition to [our high calibre staff], the consultants provided a methodology'.

> The consultants really assisted, they helped with templates.

The project was driven by technical considerations and focused on meeting the go-live date. Organisational issues were marginalised, although there were many communication and change management initiatives, e.g. newsletters, a focus on key users, certificates, presentations, user testing celebrations, a website as an introductory mechanism to familiarise staff with consultants, badges to identify team members to the organisation, and go-live countdown days displayed on staff PCs.

> There were only 1½ people on the people side [and 34 on the technical side].

Apart from the CEO and project sponsor making regular visits to the project room, there was little other direct managerial support. This became an issue when some staff expressed the concern that they were going to lose their jobs. 'Management were not prepared to say that they had made a decision ... I had to check with the CEO myself and then tell people their jobs were safe'. This interviewee felt that senior management lost an opportunity because they did not put their 'name to the newsletters'.

> The ERP is a big program, it sits everywhere. Y2K was only one issue. Top management knew about it, but was not well communicated to staff [who needed to know] facts about consultants, team, implementation, etc. Not all managers communicated to staff...

'It was seen as a technical implementation [to reduce systems] rather than process improvement'. An action item was minuted for one of the steering committee members to liaise with a re-engineering project that had commenced prior to the ERP implementation. Despite this, not one of the interviewees mentioned the re-engineering project. The CEO himself said, 'this was never a re-engineering project'.

Stage 2

Steering committee minute:

RP noted the continued efforts by project team members in addressing Stage 1 issues resulted in a drain upon the Stage 2 resources. He opinioned this was due to TechMedia not taking ownership of the [ERP] system, not encouraging user understanding and requiring the project team to continue to fill the support void. This point was disputed by several ... committee members.

By Stage 2, 'the team was tired' and the issue of ownership was causing many difficulties. A project team member was in charge of delivering answers to questions from the shop floor but after it was implemented in Stage 1, 'she couldn't do it over and over again and still work on Stage 2. She needed a meeting with the director and associate director to communicate that after Stage 1 that key users were the people to contact not her. She needed to tell them to take ownership'.

> Management was reluctant to get involved. It was sometimes difficult to get sign-offs for software design changes and almost impossible to get a commitment for changes in the business process. As a result, too many customisations were made to the ERP package to make it fit with the existing business processes.

Sometimes, there was ownership, but accountability for the results may not have been thoroughly considered. It was reported that during Stage 2, some business users were pressured to sign off on design specifications that would not even provide the functionality of the existing system. When queried about this, one of the business unit managers said, 'At the time, it seemed like the way to go'.

A senior manager was particularly difficult. He would always state that he was committed to the project, but he allowed the ERP to be implemented in his area with a number of shortcomings. Because of his seniority, the project team was unable to force issues with him. He disagreed with some of the early design decisions and felt that he was 'being shouted down by the other members of the steering committee'. He withdrew psychologically from the process and eventually left. His replacement could not understand why processes in his area had not been reengineered and started to blame the system, which was significantly slower than expected. The CEO was briefed on the issue, but it did not become clear to the CEO that he had to intervene until it was too late.

User acceptance was delayed because the technical problem took almost nine months to resolve. However, the entrenched culture of TechMedia appears to have compounded problems. Two similar functions were performed in different functional groups but 'the groups didn't communicate to realise that they should have integrated ... and what happens now is that information has to be double entered'. A number of major re-engineering initiatives in one division were not going well and senior managers concluded that too much was being attempted at one time and some initiatives had to be deferred. The board accepted management's assessment of the situation but were quite unhappy that the technical functionality had been implemented without the benefits being delivered.

Minutes of the steering committee noted two major risk items for the entire length of the project. Mitigating actions were never taken, both risks eventuated, and the predicted difficulties occurred.

The first of the major risk items was the likely loss of key staff. No action was taken to find an interesting post-project role for them. A number of interviewees believed that some managers felt threatened by how much they would have to pay to retain their ERP expertise, so they allowed budget restraints to prevent them from dealing with the issue. Key staff left as predicted and when enhancements were attempted after the project, the absence of people who understood business process interrelationships 'doubled or trebled the time needed to make changes'.

The second risk is less visible but probably more costly to the organisation. Every steering committee meeting minuted that an operating division manager should liaise with the other two divisional managers and explore further opportunities. The operating division manager had identified $500 000 of savings in his area and the steering committee thought it was quite likely that the same opportunity existed in the other two divisions. This action item was never closed and there is no evidence that it ever happened. It looks as if the other two managers did not want to participate. The option to realise another $1 000 000 of benefits was lost.

Outcomes

Stage 1, to replace the financial systems to meet Y2K compliance, was an outstanding success and the project is used as a reference site by the vendor. 'It was the fastest implementation in the world with not even one day of trading lost ... Testing had been done; training had been done. The customer service people went live with the new system straightaway'.

The outcomes from Stage 2, implementing the other requirements, could be interpreted as a success, a partial success or a failure depending on the expectations. The CEO believes it was a success despite not delivering on all the benefits presented to the board. For him, the board presentation was just a hurdle to be jumped because his objective was listing on the stock exchange and he never thought of the project as a re-engineering project to realise financial benefits.

> The CEO wanted the project to be the catalyst for the organisation to be more vibrant. Managers didn't see the vision. Only three years down the track do people see it.

One manager commented, 'We have not yet realised business intelligence available from consolidation ... one of the main benefits looked for by the organisation'.

There has been a 100% adoption rate but some users report it was not a success because there is less functionality than before, there should have been more integration and in one department, administrative work takes 3–4 times longer and has reduced the ability to do productive work. Some believe last year's decline in profit could be directly attributed to the lost productivity instituted by the new system (the available data could neither confirm nor disprove this statement). It seems the adoption rate is a meaningless measure because it is an enterprise-wide system; users have little choice.

In terms of the business case presented to the board, some benefits have definitely been delivered. A post-implementation review, 2½ years after the implementation, has found around half of the promised benefits are expected to be realised. This is less than a quarter of the benefits suggested in the business case. The project sponsor said:

> Benefits have been delivered, but there could have been more. Time was too short, and making up for mistakes has taken too much time ... the project has a 60 percent success rate, but wasn't 80 percent because of [ownership].

From a strategic perspective, the organisation is lagging behind its original vision. At the time of writing, the process management features in the newest release of the ERP had been available for 18 months, but the organisation had no immediate plans[2] to implement it. A manager explained it as partly 'a matter of timing' but to some extent, 'the organisation has lost some confidence in its ability to realise the benefits ... and [in the absence of benefits] it is hard to justify the additional cost'.

In hindsight, many managers realised they had convinced themselves that if their many different systems were eliminated and replaced with 'an all-singing

---

2 In fact, the many customisations made during the implementation made it very difficult to apply software patches to fix bugs without considering upgrading the latest version.

all-dancing system', the benefits would just get delivered automatically. All the interviewees raised the issue of ownership and many were very strong in stressing this point. A senior manager said:

> You can't overemphasise internal management commitment... you can measure success by the passion of management.

> Lack of ownership was the key factor in the non-delivery of expected benefits. Ownership is critical. If you have real owners, projects will be successful.

Appendix B

SkyHigh Investments[1]

Top Management Support

Background

SkyHigh Property Investments is a subsidiary of a major investment bank. SkyHigh's success depends on its ability to recognise, purchase, and manage quality commercial properties for its clients and its ability to attract quality tenants and keep them happy while controlling expenses. They also need to provide their investors with timely information.

The CEO of SkyHigh had trebled the size of the company in only four years. The organisation originally managed only a handful of buildings for a superannuation fund, but by the end of 2000, it had over 100 major properties, many thousands of tenants, and thousands of investors across a number of investment vehicles. The enormous growth in complexity was imposing operational stresses on the organisation. The CEO was acutely aware of it because two companies in the industry had recently lost market share because poor operational systems had undermined investor confidence.

Project Initiation

The CEO knew he had to put in place the right infrastructure to maintain and support the growth. He was impressed with the way a manager in the parent company had managed the company's Y2K compliance and recruited him in early

---

1 https://www.researchonline.mq.edu.au/vital/access/services/Download/mq:10163/SOURCE2?view=true.

Project Benefit Realisation and Project Management: The 6Q Governance Approach,
First Edition. Raymond C. Young and Vedran Zerjav.
© 2022 John Wiley & Sons Ltd. Published 2022 by John Wiley & Sons Ltd.

2000 to be the new chief operating officer (COO) of SkyHigh. The CEO also recruited Paul Major from a competitor to work with the COO. Paul had an accounting background similar to the COO and he had also worked extensively in the industry. Paul had recently been involved with the implementation of an ERP system in another firm.

Paul started by interviewing all 100 staff to identify the major issues. According to Paul, SkyHigh had 'bad systems, bad processes, bad support structure, and the wrong mix of staff'. He summarised his findings and presented them initially to the COO and then to the senior management team.

At the senior management presentation, it became apparent to the CEO that most of the senior managers had not read Paul's briefing paper and they were not able to fully participate in the discussion. The CEO decided the best way forward was for Paul to further summarise his work and identify the major projects that needed to be undertaken. The CEO was concerned because the group had grown rapidly through mergers and acquisitions producing individual units of excellence but with limited cohesion across the group.

Paul identified seven major projects that needed to be undertaken, one of which was to replace or enhance the existing management and accounting systems. There were three related projects: (i) appointing more experienced accountants, (ii) implementing a Web strategy, and (iii) integrating with forward planning systems. The other three projects are not directly relevant to this case.

Establishing Project Governance Structure

The COO took the responsibility to oversee the seven projects and Paul managed their implementation. These were the major projects. So, Paul worked with the COO to identify the key stakeholders and develop an overall governance structure.

SkyHigh had an official ICT project methodology and all project managers went through project management training. The COO had been through the training and had used the methodology in the Y2K project. However, neither Paul nor the COO felt the official ICT methodology had much influence on the governance structure because 'it's all just common sense really'.

The head of SkyHigh IT, concurred by saying 'when users have enough project management experience we don't need to get as very heavily involved in the business re-engineering and can focus on implementing the technology. In projects with less business unit ownership, we appoint much higher-level project managers with more of a business analysis focus who are able to manage the business change as well as the technology implementation'. He added, 'it's better when the business takes ownership . . . in this project, once the scope of the work was agreed and the governance put in place, I was able to monitor its progress and the business's satisfaction through the steering committee'.

It appears that it is a common practice within SkyHigh to customise a unique governance structure to reflect the individual considerations of different stakeholders. Paul said of the governance structure of a more current project, 'it's taken me weeks to reach this understanding of the project'.

The final governance structure reflected what the COO and Paul thought would be most effective, taking into account their understanding of the people, interpersonal dynamics, and standard processes within SkyHigh. The steering committee consisted of only the COO, Paul, the head of IT, and the head of finance.

> We asked ourselves whether it was possible to manage the project with a steering committee with all the stakeholders, decided it was not, and decided to keep the steering committee small.

A group of key stakeholders from finance, the business and IT formed the systems working group. Their role was to develop the list of requirements and select a new system based on their criteria. A memo was sent to formally advise them of the project and seek their participation. They would have been aware of the project through Paul's original interviews and unofficial feedback from their managers.

Paul eventually created a control group between the two groups at the time when the new system was implemented, after the requirements analysis and after funding was approved. It consisted of himself, Neville Sergeant (an IT group project manager), and Lisa Smith (a project officer to manage the technical and administrative aspects of project management such as the creation of Gantt charts, identifying resource conflicts, updating progress, and monitoring slippage).

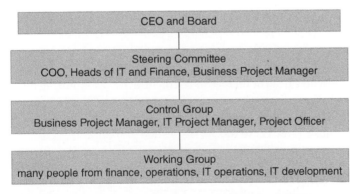

Figure B.1 Project governance structure. *Source:* Young, 2005.

Requirements Analysis

The property management and accounting systems project commenced in Jan 2001 with Paul interviewing each of the key stakeholders to determine what the new system should do. Paul felt it was important not to restrict the list of requirements but made sure it was clarified whether a requirement was critical, nice to have or a wishlist item. In making this distinction, Paul made sure he understood the underlying business process.

To manage other commitments, each interview was restricted to about an hour (they ranged from half an hour to up to two hours) and additional interviews were scheduled as necessary. In practice, it took about four interviews per stakeholder. This is quite a significant commitment but Paul reported that, in general, people just made the time and where there was a difficulty, they would schedule interviews for early in the morning, late at night, or book, say one week in advance.

Once these initial interviews were finished, a 30-page document was prepared, summarising all the requirements. Paul added his prioritisation of the requirements and sent it back to the stakeholders for confirmation.

Paul said, 'I was very conscious that no package would be able to do everything and tried my best to manage expectations'.

A Useful Delay

Towards the end of this process, around the end of March 2001, discussions were being held with contacts in the industry to identify potential vendors and initial discussions were held with ten to let them know of the process being followed. All interviewees felt the project would have continued and probably would have been implemented by 1 July 2001 but a number of incidents changed the priority of the project.

- A number of mergers had been finalised to substantially increase the size of the business and the acquisitions had to be integrated as a matter of urgency.
- A website needed to be developed to present a consistent image to the investor public.
- The COO appointed a new head of finance to restructure the accounting department. Many staff changes and many new appointments followed.
- There were unfavourable internal and external audits requiring dedicated business resources to resolve (NB: There was no suggestion of fraud).
- Paul got married in August and took leave until the end of September.

The intention was to implement a new chart of accounts; at the same time, a new system was implemented. This was only practical at the start of the financial year

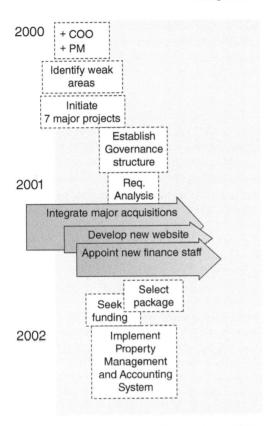

Figure B.2   Project timeline. *Source:* Young, 2005.

on the first of July, so the above incidents made it impossible for 2001. The decision was made to delay the project and aim for 1 July 2002.

The business had built up expectations for a new system and had to endure an inadequate system with all its workarounds for an additional 12 months. The need for a better system became firmly fixed in almost every one's mind.

Package Selection: Understand Workarounds and Trade-offs

The project resumed in a kick-off meeting around October 2001. The COO introduced the project himself and said:

This is the biggest project that we will do this year.

The significance of this sentence was not lost on the attendees of this first meeting. The company had grown substantially and seen a lot of big deals yet 'this project was more important [than the last major acquisition]'. It was obvious to everyone that the company was taking this seriously. The COO had considerable personal power in the organisation, and he was supported by the CEO. Each project team member had salary bonuses added to their annual review with criteria tied to project performance objectives.

The first key task was to compare potential replacement systems with the requirements document. The working group was quite large and crossed many functional areas. The potential for conflicts of interest was quite significant yet not one of the interviewees mentioned the issue except for the difference in preferences between IT and the business.

> We narrowed down the selection [quite quickly] to the two with the business clearly favouring one while IT preferred the other. We had to work through issue by issue before IT signed off on the final package.

Paul was primarily a business user and recognised his weakness in matters IT. He could have forced the issue but he chose not to, recognising the important contribution IT had to make. 'Neville is the most knowledgeable IT person I have ever met'.

> He took the time to explain to me the implications of various decisions [in business language].

Neville said, 'we had a standing joke at the beginning of each day 'Paul, time for your tutorial'. Neville added that he is normally not good at this, but

> Paul wanted to understand and made the effort.

Paul did not accept that Neville is normally not good at explaining technical issues and the implication is that business managers tend not to make enough effort to understand. As each issue was raised Paul would systematically interrogate with the following set of questions: (i) Why it is essential? (ii) What happens if it doesn't? (iii) What happens if it breaches our standards? (iv) How can we work around this?

The governance structure and the attitude of the decision makers clearly made a difference.

- Sometimes the workaround was that senior management would formally acknowledge acceptance of certain risks, risks that the individual department could not accept.[2] In these cases, Paul would talk to the COO (who would, in turn, talk to others as necessary), get approval in principle, and formalise this with an e-mail from the individual department noting the risk and a response from the COO accepting the risk.

- In other cases, IT would develop the workaround solution[3] even if it would cost them more time operationally. Their attitude was 'if this is what the business wants, and we can find a way to work around the problem, we shouldn't stand in the way even if it creates more work for us'.

This issue highlights that the general approach followed was to discuss and try to convince. Interviewees commented:

> We always felt like we were heard . . . there were lots of meetings and sometimes they went over time [to discuss issues that were important to us] . . . We accepted the pressure to meet the deadline, but we were never pressured into accepting something we couldn't accept.

However, there were also times a less conciliatory approach was followed. The COO admitted to saying 'we don't want to hear this sort of argument' and it was reported that Paul would occasionally put a user in his place based on his knowledge of the overall business process and not allow someone to insist on inefficient current practice. The final choice was formally signed off by each of the ten key stakeholders and the overall sense was that everybody was satisfied because all the issues had been properly resolved. 'Paul and the COO really earned their money'.

By the time the package was selected, there was a good understanding of what it could and it could not do. The limitations were understood, workarounds had been developed and the risks were acknowledged and accepted by the appropriate stakeholder.

The entire senior management team had attended a separate strategic briefing session with the potential vendors to evaluate their suitability as a long-term partner of the business. The successful vendor commented, 'I knew from the moment I walked in that this project would be successful' [because of the level of understanding and support from the senior management team].

All interviewees commented on the importance of the work done in the first 12 months to the eventual success of the project. The length of time the project had been in gestation psychologically prepared the business and established their

---

2 E.g. no automatic back-ups (running the risk of failure and downtime for operations if the backup had to be rerun during the day).

3 E.g. manual back-ups.

will to change. The commitment of the project team was very high, but the COO knew he was taking a gamble because 'we are going to change to suit the system'. All existing business processes were changing.

> You look at the culture, the willingness to change, and the perception of how big an issue you're fixing . . . it took us a lot of time, a lot of communication, and a lot of discussions to reach this point. . . there was a desire to improve, we knew what we wanted from day one and we had everyone's buy in . . . I knew the type of people I had and how far I could push them . . . in a different organisation, this would have taken twice as long [e.g. Government].

Obtain Funding

Concurrent with this effort, board approval for the project was being sought. However, it was initially rejected for budgetary reasons. The COO said he and the CEO tried again and again because 'It's a matter of passion and what you believe [needs to be done]'.

The COO and the CEO spent significant time lobbying the board both individually and collectively to seek approval for the project. The input of the individual board members was actively solicited to allow them the chance to contribute to the project and feel that it was also theirs.

> As a senior manager, it is important to have no surprises. We spoke to the board members individually to convince them of the need for the project and to allow them the chance to contribute to the project and feel that it was also theirs.

Project Implementation: Monitoring and Managing Risks

Funding was approved and the formal project implementation commenced in January 2002.

An experienced project manager (Lisa Smith) was appointed to a project officer role to assist Paul with the day-to-day running of the project. This was to allow Paul the chance to focus on the strategic points of the project and integrating with the other six key projects in particular. There was an allowance for three other contract staff in the budget. SkyHigh is a lean organisation and the availability of this level of resources was another clear signal of the level of senior management support.

Because of the extensive communication during the evaluation process, the contract was signed with the vendor with minor delays. Paul, Lisa, and the vendor had developed a detailed project plan and the project was subsequently monitored very closely against both the project plan and a risk management plan.

There was a lot of informal communication daily to ensure issues were being raised at the regular weekly meetings of the project teams and the systems working group. Lisa handled most of the day-to-day administrative tasks while Paul focused on one-on-one conversation, making sure he knew what the real issues were. Paul lent a hand with work when necessary and kept all the key players informed of progress.

The steering committee met fortnightly on a formal basis mainly to ensure the risks were being managed and to check that the benefits were still likely to be realised. According to the IT department, the project governance (structure and control mechanisms) is tailored to fit each project, but risk management is a mandatory component. Paul said, 'The steering committee focused more on the things we knew the system couldn't do and the workarounds [i.e. the risks]'.

Interviewees described this period as being almost incident-free albeit very intense with very long hours being worked. They had their regular job that included year-end reporting and also had to clean/correct their data from the various existing systems and preparing it for upload into the new system.

> They worked weekends.

Some in the finance department were also new to their jobs. No one described any difficulties except for one incident. Paul recalls that at one steering committee meeting, the COO noted to one of the other managers, 'you're falling behind' and the next day, the problem was resolved. The COO recalls saying something stronger in this incident:

> If you can't do it in time, we'll find someone who can.

During the post-implementation review, one project member commented that the only improvement he could recommend was to have found out earlier that some data would have to be manually entered rather than automatically uploaded as originally planned. In this particular incident, both Paul and the project officer ended up manually helping to load the data.

A detailed review of the minutes of the various meetings reveals that unexpected incidents did occur with the potential to delay the project. The implications on the overall project were noted, options for resolution were discussed, and they were assigned to specific people for resolution. All the incidents were resolved in two weeks or less. Interviewees attributed this to the formal and informal governance structure allowing rapid identification of issues, the rapid escalation of issues, the high level of senior management ownership (willing to take decisions and accept risks) and the type of people on the project (driven to achieve).

In the final go/no go meeting, everyone signed off. Paul commented that this was the riskiest part of the project from his perspective because he could not control the quality of work. It depended entirely on the owners of the data, how well they had resolved inconsistencies with the existing data, and how accurately it had been loaded into the new system.

It was made clear to everyone that they did not have to sign and go live with the new system for their particular module and that a workaround would be found (but, of course, it would be inconvenient to them because they would have to enter their data twice for an additional month of parallel running until the problem was solved).

The COO said, 'The people who did this work are still with us now'. His implication is that the way you manage this risk is by assigning tasks to people who care and would be accountable for the result. Both the COO and Paul commented that it was important for the business rather than IT to have managed the project. Paul's comment is particularly salient in the light of his experience with a similar but less successful ERP implementation. (That project had an IT/technical focus).

Outcomes

A year after the implementation, all the interviewees consider the project a success. Not long after the go-live dates, the whole project team was taken out to dinner to acknowledge their work. One member of the team said:

> It was something I was proud to have been part of.

In project management terms, it is a clear success because it came in on time and below budget and worked.

The business benefits had never been expressed in easily quantifiable terms.[4] The subjective assessment of the COO is that most of the business requirements were met (96 to 97%). He points to the following indicators of success: 'A much higher level of data integrity. The accounting close-off on 31 December was a far

---

4 There was perhaps a board discussion on the expected business benefits, but apart from the COO, it was not mentioned, and the COO was not at liberty to give general access to board minutes.

cleaner process (and has probably overcome the problems that caused the poor audit result in 2001). The processes are maturing'.

The COO mentioned some minor problems[5] but his overall feeling was 'it was amazing there weren't more problems'. Interviewees generally had to think quite hard to think of things that could have been improved. The only exception to this was the issue of time pressure.

The post-implementation is still being conducted and more lessons may emerge, but there is almost no question that the project succeeded and was perhaps completely successful in implementing a system that would support the growth of the company.

---

5 There were some data issues. Not all the processes were updated. Some were still pushing the old way of doing things, but the COO felt this was part of an ongoing continuous improvement process and was not overly concerned. He also accepted that some of the policies and procedures were not finalised with documentation.

Appendix C

The Agency[1]

Succeeding Against the Odds

Background

The Agency is a public sector organisation providing scientifically derived information services. It is recognised as a world leader in its field and Agency personnel lead the international coordination to further the science.

The Agency recognised the potential of computing from its earliest days. It has successfully used computers since the early 1950s when the Agency was closely connected with the third or fourth modern computer ever built. The Agency staff are constantly at the leading edge of technology (e.g. high-end supercomputer projects) but their focus has always been to improve operations rather than simply exploring the technology. Their business achievements include the reduction of staff numbers by over 30% (over the period 1976–2003) while simultaneously providing more information at higher levels of accuracy.

The Agency has a track record of success in its core business and deserves its reputation for doing ICT better than most. However, Agency staff have different opinions[2] on what are the crucial elements that lead to its consistent success. This case study has been undertaken to try to clarify which aspects of the Agency's ICT practices have the most impact on success. It is a difficult nonstrategic project undertaken in a nonoperational department of the Agency. The case is particularly interesting because the relative success of the project can be attributed mostly to the support of two senior managers. This is particularly significant

---

1 https://www.researchonline.mq.edu.au/vital/access/services/Download/mq:10165/SOURCE2?view=true.
2 Because there were so many different opinions, and even though many direct quotes are used, the presentation of the case necessarily reflects the understanding of the author and is not necessarily the view of the Agency.

Project Benefit Realisation and Project Management: The 6Q Governance Approach,
First Edition. Raymond C. Young and Vedran Zerjav.
© 2022 John Wiley & Sons Ltd. Published 2022 by John Wiley & Sons Ltd.

because although there were some good staff, the project overall was quite under-staffed and not particularly well planned.

The Finance Department

Most Agency staff outside the Finance Department have professional or scientific backgrounds. The work is very specialised and the quality of their work is reflected in the observation that 'once you join the Agency, you never want to leave'. In contrast, most of the Agency's finance staff have clerical backgrounds. They also have long average tenures (of around 15 years or more), but in the finance depart-ment, this was not for positive reasons:

> We were one of two organisations [in Australia] who used the previous computer system . . . and we were cash-based which is significantly easier than accrual accounting. After a few years, that was all you knew, and you couldn't get a job in another organisation.

Most Agency staff considered the finance department something of an inconven-ience and largely irrelevant to the operational business of the Agency. This position was somewhat justified because the Agency had reached world-class standards through a focus on other 'core' areas of its business over many decades. However, they were being forced to broaden their focus to make a bigger role for financial reporting. Other public sector agencies were not as focused as the Agency and many were very inefficient. The government was demanding more financial information from all public sector organisations and the Agency had no choice but to comply.

The Agency Head was not against information that supported requests for funding or helped the Agency to strategically allocate resources but he was strongly opposed to the 'management fads . . . being imposed' where such activities impacted on the overall objectives of the agency. The Agency operated in an era of government cost cutting and the Agency Head had 'quite fixed ideas' on the information required. The finance department was to comply with the government requirements, but the Agency had a separate Executive branch to produce management information. It was effective for the Agency, but the finance department felt neglected.

> We're not second-class citizens, we're third- or fourth-class.

Project Failures Within Finance Branch

The Finance Branch were left alone to do their job. Top management were reactive as opposed to proactive to their activities and paid attention only if there was negative publicity from the government, the national audit office, or some other outside source.

The Finance Branch implemented five separate information systems (AP, AR, GL, Inventory, and HR). They worked, but they were not integrated, and significant manual intervention was required to produce the statutory financial reports. Budgets were managed and audits were passed without qualifying comments but there was significant unease about the quality of the information provided. They frequently had to 'apologise to the executive for being late with reports and had to delay ministerial requests for information'. The feeling was they were only just avoiding external scrutiny and they were running the risk of incurring significant embarrassment. This would have been unacceptable given that the Agency was seeking Executive Agency status in an effort to gain more control over their destiny.

The situation was compounded by a number of information system project failures or perceived failures:

- The Agency was tolerating a 'bug-ridden' inventory system 'inflicted' on the organisation in 1983. It had been customised to perform financial reporting but after almost 20 years, 'there was still no end in sight' to how much more would have to be spent to either fix or enhance the system.
- There was a large cost overrun followed by the outright failure in the implementation of a General Ledger system in 1993.
- The Management Information System (MIS) initiated through the Executive Branch 'failed' to deliver high-level information (to complement the financial reports required by the government).

All 10 members of the senior management team had sat on the steering committee for the MIS but it proceeded slowly, with many initiatives and proposals being rejected, and it took almost 10 years to produce the system. It was a technical success (providing expenditure and revenue reports against budget for a wide range of line managers) and was possibly the first data warehouse ever developed in Australia. It was widely used within the Agency, but because it did not meet the high-level reporting expectations, e.g. performance reporting, its successes were not acknowledged.

This situation was extremely frustrating to the senior management. The Agency has consistently succeeded with high-risk supercomputers and leading-edge ICT projects. So, it was difficult for them to accept 'failures' with 'easy systems'. The feeling was such that 'top management despaired that anything could ever be done right in the Finance Branch'. The situation was tolerated because the shortcomings in the finance department did not affect the ability of the Agency to reach and improve world-class standards in its core business.

Accrual Accounting and Other Drivers for Change

Around 1997, several government initiatives changed the situation. The most significant was a directive for accrual accounting to be introduced by 1999/2000. A second directive was to standardise computer systems across the whole of government. These two drivers had serious implications because Agency financial management systems were neither standard nor capable of handling accrual accounting.

A new Finance Director was appointed. She had some financial background, but she was unfamiliar with how to get things done within the Agency. She was the first senior manager to be appointed from outside the Agency in many years and the only female senior manager. She had a difficult decision to make and it was made more difficult when her peers would stop her in the corridor and ask her when she would make a decision about the financial system. She realised:

> It was a tough project . . . on the bottom end of the totem pole.

> I didn't get a lot of time with the Agency Head . . . I knew I would not be able to change them [and get their support] . . . so I had to reach the point where I was convinced the project had to succeed and then . . . I would do it with them or without them.

The new Finance Director reports that it took her some time to get to grips with the issue. She 'would not be panicked into making a decision' and needed to reach the point where 'there was real ownership' on her part. The problem was that there was 'a serious lack of skills' in the finance department (who did not understand accrual accounting) and there were conflicting and strongly held opinions for the best way forward.

Henry Tell, her senior financial officer, was of the strong opinion that the existing system should be upgraded. However, this option would result in a nonstandard system and was risky because 'so much money had already been spent on the existing system without providing a satisfactory system'.

Henry had commissioned a consultant to explore the issues, but the computer support manager felt strongly that the main recommendation (to develop a new system in-house) was beyond the capacity of the Agency. Mark Black, the project manager of the 'failed' MIS project, commented in writing that the consultant's advice to develop a new system was more complicated than might first appear.

The Right Advice and the Right People

The new Finance Director needed advice she could trust upon. She turned to the chairman of the MIS steering committee who had been instrumental in appointing her to the Agency. The chairman firstly helped by reaffirming his confidence that she had the ability to drive the initiative through. Then he helped her see that the organisation's poor perception of Mark Black was largely unjustified. This was confirmed very strongly by Mark's former manager's high opinion of him and his confidence that '[Mark] shouldn't have to wear this criticism'.

It was an important clarification for her that the MIS did not fail technically but because of a lack of sponsorship. The MIS project 'failed' because no one from the senior management had taken the risk of championing the project even though the Agency Head himself had initiated it. She realised that she faced the same risks but all that was needed was 'more direction [at a senior level]'.

This played to her strengths as a hands-on manager but still there was a considerable personal as well as organisational risk in undertaking such a project. Before she could commit to the project, she 'needed the right people to bring along with her'. The Finance Director describes her role in this way:

> You have to stand between the CEO and the workers . . . and have enough faith to believe it is going to work . . . and take the flack (to protect the team) . . . but you have to let the team know something of what is going on, so they are aware of the sensitivities.

> You have to have confidence in your team and back them. On occasions (when they make mistakes), you may bruise them up a little (but not to the extent that they lose confidence).

> I've learned if you back staff . . . they will rise to the occasion . . . (unless they are clearly unsuitable, in which case you need to find someone else).

With respect to this last point, the Finance Director said that it takes her time to commit this level of support to her staff.

Initiating Project – Request Funding

The first overt sign that the Finance Director was committing to the project came almost a year later in April 1998 when she formally requested funding from the MIS steering committee. The request was to undertake the first of the consultant's recommendations (to determine the data requirements and functionality needed) with the intention to replace/upgrade/outsource the Agency's financial systems.

The MIS steering committee approved the decision to go ahead. A small project team including Mark Black and Henry Tell (the senior financial officer) was assembled to determine the data requirements and functional specifications of a new financial management system. This was undertaken over the second half of 1998. The technical strength of the Agency was shown in the latitude and confidence given to the staff and their technical background. Consultants were not used and staff were given the time to 'read a couple of textbooks' and work out what to do. Key stakeholders down to operational levels were interviewed and business processes were mapped along with dependencies and planned enhancements.

The project team members made the comment that it helped enormously when the Finance Director made it clear she was 'absolutely committed to the project'. They were aware of the poor esteem in which the Finance Branch was held.

Winning Allies

The request to initiate the replacement system was a positive move, but it must also have been received with some angst at the senior management level. Funding had been set aside for the upgrade of the financial systems, but (i) senior management and the Agency Head, in particular, did not like to 'waste money' on financial systems; (ii) the track record in Finance Branch was poor; and (iii) there was no reason to expect the Agency would not be the next 'horror' story given the bad experience of their colleagues in other public sector agencies.

The Finance Director said that although no one ever confronted her with these fears directly, her peers in the corridor were 'quick to say how the existing system was inadequate'. It had become apparent that the major risk was not technical but the potential lack of support. The Finance Director was still relatively new to the Agency and was still learning the rules for how things were done.

Another person became 'very important to her' – Brian Minister. Brian had served for several years as the Agency Head's right-hand man in respect of budget matters, had Finance Branch experience, and is an organisation troubleshooter often assigned to the most difficult assignments. His high esteem is reflected in his subsequent promotion to a director position and, at the time, was perceived to understand the Agency Head well enough to represent his opinion.

When Brian openly supported the project by joining the finance system steering committee, it sent a clear signal to the rest of the organisation.

Overcoming Dissension in the Ranks

The Finance Director worked steadily at increasing support and putting the governance structure in place. In the meantime, the working group reported to the steering committee that the option to outsource or to develop a new application was not viable. The main reasons were that there was nothing to outsource and there was not enough time to develop a system in a house that would be capable of full accrual accounting reporting by July 2001.

The steering committee accepted this recommendation and agreed that a mini-scoping exercise be undertaken to evaluate various commercial off-the-shelf financial packages. The option of upgrading the existing system was also to be evaluated.

These options reflected a major issue that had been simmering in the background. Mark Black (the project manager of the technically successful but poorly perceived MIS project) and Henry Tell (the champion of the existing bug-ridden system) had very different and very strong opinions of what the final solution should be. Mark felt that the system should not only report financial data for external purposes but also provide information for internal purposes. Henry opposed any additional complexity and believed that the best option was to enhance the existing system. The issue was initially resolved when the steering committee supported Mark's view. Henry however did not agree with this decision and continued to actively promote his views. Some of the documented incidents suggest a high level of tension:

- Henry refused to participate in the formal evaluation of some packages despite direct instructions in writing to attend.
- A file note recorded a conversation where Henry accused a colleague of interfering and ended with verbal abuse.
- Henry lodged a formal accusation of impropriety (in the selection process for packages to be evaluated) against the Finance Director.

During this period, there was a great deal of corridor talk. There were undercurrents at one level suggesting the viability of the Agency was being threatened because the new system would not work and at another level, by suggesting that they were being unpatriotic by not supporting a local developer. Senior managers were never confronted directly with these misgivings but they 'picked it up in the corridors'.

The Finance Director dealt with the undermining rumours by calling meetings to air issues 'not knowing what would come out'. In the case of the formal accusation, the Finance Director arranged for the two accusers to have the chance to present their grievances directly to the MIS steering committee and then for the Agency to seek an external audit to clear itself of the accusation.

This period ended when both Henry and another opposer took leave and, finally, left the Agency. The issue was managed sensitively to show respect to all parties and, while difficult to handle at the time, became relatively unimportant in the grand scheme of things. However, the resolution of the issue delayed the project by six months.

A Cautious Phased Approach – Phase 1: Proof of Concept

The working group recommended that the Agency purchase SAP. The recommendation was made because the Agency, although small in transaction volume, had many complex transactions 'similar to a merchant bank . . . [and] needed rich functionality'.

The recommendation was logical, well justified, and documented but the unspoken feeling in the MIS steering committee was 'that other government agencies had tried and failed with SAP . . . why would we be any different?' The recommendation was even harder to accept because the expected budget was a relatively large sum for the under-funded Agency.

The MIS steering committee approved[3] the project but they directed a cautious route. It was very important for them to avoid being in the next disaster. The experience of the Agency is that projects go wrong when insufficient thought has gone into the preparation and they wanted to undertake (i) an initial proof-of-concept phase to give them the option of terminating the project before committing to (ii) a full purchase and implementation.

This proof-of-concept phase was undertaken primarily by Mark Black. Mark believed that the project risks, in particular, the acceptance of the system by users, would be minimised by making as few changes to the existing business process as possible. His task was to prove SAP could be implemented in the Agency with all the required functionality with minimal customisation and minimal changes to the business processes. It is significant to note that the decision to defer process changes in some ways conflicts with Mark's own evaluation of SAP[4] and it is possible that in focusing on tactical issues, the long-term benefits were compromised.

Mark spent the next six months (January–July 2000) with SAP consultants and the software manual. Mark reports that the SAP consultants were very surprised to see the level of detail he wanted to explore before validating each functional

---

3 It is difficult to know exactly what tipped the balance to proceed with SAP, but three factors stand out. Firstly, the government. The Agency was one of the last to provide full accrual accounting and the Agency Head was getting more and more pressure to comply. Secondly, the strength of conviction of the project team. Mark Black says, 'I knew why we had made our recommendation and I just stuck to my guns'. The final factor is probably the quality of the team. 'I'd like to think that our executive would be smart enough not to do it if they didn't think they had the right people'. The Finance Director was told, 'Mark can do it'.

4 The mini-scoping evaluation stated: 'SAP R/3 will require considerable cultural change in current attitudes to processing financial transactions. More care will be enforced by SAP R/3 . . . a reference site identified this is causing initial resentment. The point to make, however, is that the Agency's current approach to coding and entering transactions is inadequate'.

requirement. He says that 'the consultants were used having their clients accept their advice at face value', but 'they can't know your business as well as you'. He valued the consulting advice but did not over-rely on it.

> When consultants tell you they know how to configure their software, they mean they know a way of configuring the software, but they may not know the best way.

> When consultants say it can't be done, they mean they don't know how to do it.

Mark feels that the Agency made a very good decision by appointing SAP as the prime contractor with overall responsibility for all other contractors. SAP introduced an implementation consultant but when the consultant said 'it can't be done, it will have to be a customisation', Mark would sometimes call on SAP to provide an expert to work out if it actually could be done. Mark was seldom wrong and he based this judgement call on his best understanding of how SAP worked. Mark felt the dynamics of this phase arose because consultants do not know your business, they have a conflict of interest between wanting to help the customer and making a profit and because some consultants don't know the product as well as they should.

The proof of concept was relatively trouble-free because Mark had been intimately involved in the preparing of the functional specifications. He understood the business processes well and his years of experience within the Agency prepared him to anticipate future needs. He was also aware of the major difficulties other agencies experienced.

The end of this phase produced the Blueprint, a conceptual map of how each business process could be configured in SAP.

Phase 2 – Implementation

The Blueprint proved that conceptually SAP would work. It was completed by July 2000, three months longer than originally planned. The delay was not a problem because there was no significant cost involved and the main objective had been met (i.e. to be sure, it could be implemented in the Agency environment). The implementation phase to follow was quite different because significant costs and resources were required. A detailed project plan had to be prepared to avoid budget difficulties.

The detailed project plan relied heavily on the advice of the external consultants. Mark knew he had to take detailed advice because he operated more at a conceptual level and this was not his area of strength. He said, 'I had to. We had never done it before, so how would we know?' Unfortunately, the plan that was prepared did not reflect the organisation realities at the Agency.

- The project plan was unrealistic in the Agency environment because it was too labour-intensive, with too many activities occurring in parallel.
- Insufficient time was allocated to training.

The first issue was hard to detect because it was presented as a list of activities rather than schematically. The consultants' plan required eleven full-time and four part-time project staff with most activities occurring in parallel. It recognised 'the provision of this level of support to the project is . . . one of the high-risk areas' but the extent of the risk was not fully anticipated.

It became a major issue because the consultants could not have realised the antipathy with which the Agency viewed their financial system. The consultants were on a fixed-price contract and they fully expected the activities to proceed according to the project plan. When slippages started to occur, they put enormous pressure on Mark and other project staff to complete planned tasks, even though they were already working very long hours.

The problem was that of the fifteen staff required by the plan, only seven staff were really available to make any significant contribution to the project. Mark knew better than to ask for the 'best' staff to be allocated to the project. He reports that Agency managers would incredulously ask 'you want more?' to his requests for the basic level of representation. Some project team members described themselves as a 'second eleven'. Mark defended a few colleagues, but he acknowledges 'he would have liked more senior people on the team'.

The understaffed project had almost no chance of following the consultants' project plan.

Creating an Environment to Succeed Against the Odds

To make matters worse, morale had been particularly affected by the opposition of the senior financial officer Henry Tell and one of his peers. Their views had created significant doubt about the project, doubt that had reached beyond the Finance Branch to touch many in the organisation. Many tried to avoid it by responding with a wait-and-see attitude but expectations were low and few volunteered.

The deadlines looked very shaky and the viability of the project was being challenged. Mark recalls how important the senior manager support was to him at this time and their intimate knowledge of the organisation. Brian Minister describes his role in the following way:

> To be available as a sounding board especially for the project sponsor and the project manager 'to help work out what the real problems were' and to 'explore possible solutions.

Mark would discuss issues with Brian and the Finance Director as he needed. With the issue of staff shortages, they either (i) developed strategies to ask for alternative resources, or (ii) reduced the scope of the project in a way likely to be acceptable to the board.

Both Brian and the Finance Director offered significant levels of support by making themselves easily available to discuss issues, ratifying decisions, or contributing to alternatives and helping to sell decisions. Mark particularly valued the strength of the Finance Director's commitment to the decisions made and she made sure Mark had the confidence to know that she would personally fight for them. They had by then developed a good working relationship, understood how the other thought and 'could communicate on the same wavelength'.

The Finance Director had an office on a different floor of the building, but she made a special effort to visit the project team as often as possible to 'shine on them'. She worked at creating a positive environment where people could rise to the challenges. She continued to call meetings to air issues whenever she sensed unease. She said, 'As time went on, people started to want to come on board as they could see it starting to work'. She did such a good job she even had to manage 'some jealousy because [the perception was that] the project team got all the attention'.

The feeling it was going to succeed may also have come from observing the quality of some of the key decisions. These decisions were in no way trivial and they reflected both the experience of the team and the seniority and confidence of the decision makers.

To resolve the first problem of the missed deadlines, Mark said to the consultants:

> Look there are some problems [with the schedule] . . . but I'm not going to accept that we are holding you up. [I don't know how but] we can catch up later. You've got to allow us some more time at this stage for us to come to grips with what we're trying to do . . .

The consultants rebuilt the schedule but they said, 'You're going to be in trouble later.' Mark replied, 'Maybe, but let's at least rebuild the schedule so that we are making progress rather than just panic-stricken and hitting each other over the head'.

Revising the Project Plan

The revised project plan was presented and signed off by the steering committee without difficulty. During the implementation, the team worked out how to meet the deadlines through a number of significant decisions. It is notable that no one mentions them as a major issue and it reflects on the effectiveness of communication and the transparency of the decision-making between Mark, Brian, and the Finance Director.

- The time allocated for training was inadequate and the major factor leading to the eventual delay of the project. When it was finally recognised, it was decided that external trainers were needed and Agency staff would accompany the trainers to help answer agency-specific questions. Because so few staff were available and because there was 'time up their sleeve', the go-live date was deferred to the start of the new financial year (30 June 2001) to 'give it a chance to do it properly'.
- A senior project team member said, 'I think a lot of the sign-offs are to protect the consultants rather than to add value'. In the Agency environment, it was difficult to get people to pay much attention to financial systems. So, it was decided that apart from the project plan itself, not to seek most of the sign-offs.
- We had enough experience to know that the hardware was going to work. So, 'we decided to skip the load testing'.
- It was decided not to implement the HR modules, and to defer some work till later.

One project staff member observed these decisions surprisingly and learned 'you don't have to do everything you are supposed to do'. This staff member was widely praised for his efforts in change management, yet he found he did not have to work very long hours to complete his core tasks. He explains that it is more necessary to work well with the key decision makers to scale activities to fit the resources and time available to do it.

The User Experience

There are two levels of users in the Agency, the day-to-day users and less frequent users. Their perspective of the project confirms much of the story above, but it is less clear because their expectations were so varied. The expectations may have been set too high by the project team in their attempt to get more participation:

> I initially took a wait-and-see approach, then I was sold on this all-singing-all-dancing product giving me as much control as I wanted from my budgets. [I was disappointed when] it didn't deliver.

If there is one thing I would fault on this project, it would be that expectations weren't managed properly.

Most of the day-to-day users acknowledge the system was a success and point to the successful production of the accrual accounts required by the government. The infrequent users point to the fact that suppliers get paid and parts get delivered. In all but one of the interviews, the success is acknowledged without enthusiasm because users derived no benefit from the system. Some users have a bigger workload because of the demands of accrual accounting.

Accrual accounting is conceptually much harder than cash-based accounting. The difference 'is like the gap between high school and university'. The first time most users got involved was three months before 'go-live' for training on the new system. Some users were trained a little earlier because they were asked to participate in user acceptance testing.[5] In all cases, the experience was challenging because there were so many new concepts that they had not confronted in a cash-based accounting system.

The training was organised such that they had to learn new concepts and how to operate the new systems at the same time. They mainly learned how to process the same transactions as the old system, and in this sense, was not difficult, but they also had the much more difficult task of understanding why additional information had to be recorded and the implications of not doing it properly. No one believed they mastered the inter-related tasks.

We felt like trained monkeys.

The problem was compounded because the period leading up to the go-live date was also when staff were at their busiest preparing year-end reports. There was also a delay between the training and go-live, a period of up to three months. A test system was made available for users to practise on, but few had the time or inclination to try. Users complained the system they learned was not identical to the go-live system.

This aspect of the project suffered enormously from insufficient resources. Originally, the equivalent of three full-time staff had been proposed for change management and training but in the end, there was only one person. The training was 'considered to be easy' and originally was all going to be done in-house. When the project manager eventually realised that there were insufficient resources, he then appointed a second consulting group that specialised in training.

It was a very common experience for users in the first three months after go-live to stay behind for many hours 'trying to work out what was going on'. SAP is a

---

5 Both the implementation consultants and the users challenged the quality of the user acceptance testing. The main reason given was that users didn't understand what they were accepting.

complex system and according to the training consultant, 'will never be simple'. Even printing a document required working through many keystrokes on several screens. The documentation was very good, but each person seems to have created an additional 'cheat sheet' summarising the main keystrokes to process a routine transaction. After a time, an 'official cheat sheet' was published.

It took most users around 12 months to understand variations on the routine transactions. During this period, the focus was on getting parts delivered and suppliers paid. There was almost no reporting for management activities because users had to work out 'why a transaction would work when you do it one way but not if you did it another way'. They wrestled with 'the implications of coding transactions in different ways'. The infrequent users do not have the expectation of ever understanding what they are doing and one clerical assistant said his job was to fix mistakes at the point of data entry because 'they are miscoded all the time'. He added:

> It doesn't matter because people don't use it anyway.

This last comment is very significant. Participants in the study asked that the organisation be kept anonymous because 'the jury is still out' in determining the success of the project. In this sense, the outcome is not unlike the earlier MIS project.

One of the main reasons for choosing SAP was to produce information for internal management as well as external compliance. Mark had championed 'the [internal] aspect of reporting' as far back as the MIS system in 1985. It had always been recognised that 'there will need to be a significant upgrade' to staff skills and that 'considerable cultural change was required . . . because the Agency's current approach to coding and entering transactions is inadequate'. These considerations were largely ignored in the implementation and a user commented that 'we were slow to get up to speed . . . The project team was young and keen and not afraid of change . . . but we never considered what are we going to do with people who aren't going to make it?'

Various quotes illustrate the ambivalence of the organisation to the new system: The change manager said,

> 'Within the Agency, if you don't want to do something, you don't have to . . . and training was like that. People were finding excuses not to attend'.

Users would say, 'There was not enough training', yet there is very clear evidence that much more training was delivered than ever intended. One user said, 'Resources should have been increased, not decreased after go-live' but that is, in fact, what happened. (More of the training budget was spent after go-live rather than before.)

One of the more knowledgeable users said, 'I didn't bother to go [to the user acceptance test] after the first few times . . . It was a pure waste of time because it was already cast in stone'. This user, when asked why he didn't get involved earlier, replied he offered to help by saying, 'Any time you want a hand, just give me a call'. Another user in a similar role said, 'We were invited to participate, but we didn't really understand what was going on, so we left it'. Mark Black commented that he just did not have the resources to convince more users to contribute.

Quite a few users in the same breath would say, 'It is better than what we had', and then go on to list all the faults and all the problems with the new system. On 1 July, they reported being 'excited that the system worked' and then in the months that followed, became extremely frustrated because 'we couldn't do what we could do before'. A major ongoing issue, two years after the implementation, is that users still cannot easily reconcile what is spent against their budgets.

The Outcomes

When people were asked whether the project was a success, no one gave an unqualified yes or no answer. In Agency terms, nothing remarkable happened. The project didn't fall over. It was live and working from day one with no major dramas. It met the revised go-live date and was within budget. Mark said, 'I didn't expect anyone to thank me because, in the Agency, this is normal'.

The Finance Director remembers however a very subtle indicator of appreciation. 'The Agency made a modest contribution to costs associated with the celebration lunch'. It was symbolically very meaningful because this almost never happens and to her, it indicated that the Agency recognised the achievement was out of the ordinary. One of the main concerns of the executive was to avoid as a disaster and it is significant that every member of the project team has subsequently been promoted.

In terms of the external benefits, it is now complying with accrual-based reporting requirements in a timely manner and audits have been completely unqualified. There is a high degree of confidence in the information produced.

In terms of internal benefits, however, the weakness in management reporting even two years after the implementation mars the overall success. Only one senior manager said anything positive. He appears to be a voice in the wilderness, with not even the sponsor or project manager recognising the potential with such enthusiasm.

> This is the best thing that has happened to the Agency . . . I can now justify making provisions to replace my assets and I could never do that before.

Further Reading

Markus, M.L., Axline, S., Petrie, D., and Tanis, C. (2000). Learning from adopters' experiences with ERP: problems encountered and success achieved. *J. Inf. Technol.* 15 (4): 245–265.

Index

Project Benefit Realisation and Project Management: The 6Q Governance Approach,
First Edition. Raymond C. Young and Vedran Zerjav.
© 2022 John Wiley & Sons Ltd. Published 2022 by John Wiley & Sons Ltd.